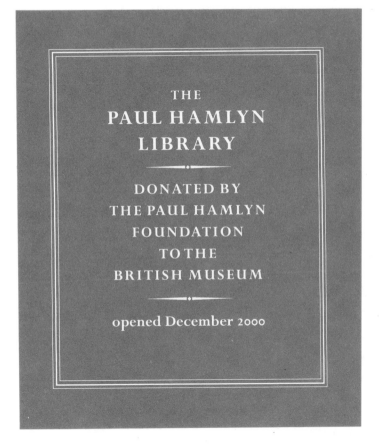

13.4.86

Far from Paradise

Far from Paradise
The Story of Man's Impact
on the Environment

John Seymour · Herbert Girardet

British Broadcasting Corporation

Published by the British Broadcasting Corporation
35 Marylebone High Street, London W1M 4AA
First published 1986
Copyright © 1986 John Seymour and Herbert Girardet

ISBN 0 563 20357 9
Typeset in 10 on 12½ pt Linotron Sabon by Ace Filmsetting Frome
Printed in Britain by Mackays of Chatham Ltd

*To the founders of the organic movement
especially Lady Eve Balfour,
and to those who are carrying on their work*

Contents

Acknowledgements

This book is a distillation of the research, thought and experiences that went into the making of the television series *Far from Paradise*. John Seymour and I decided to keep our contributions separate, writing alternate chapters, because we can thus take responsibility for our own ideas and selection of facts which we present here in a joint effort to find truth.

When talking in this book of *humankind* we have usually confined ourselves to the use of the word *man*. We would have been happier using a more neutral term but there is not one that is commonly accepted as a substitute. We are aware that *man* is an unsatisfactory word to summarise human experience and history.

Many people have helped us with our work, making the films and giving substance to the words that came to be written down in this book. We have drawn from many authors who have been credited in the bibliography at the end of this volume. We would like to say our special thanks to all the people who helped to make the television series and this book a reality: Ian Bodenham, Dr Hannes Bressler, Gudrun Guenterroth, Barbara Trottnow, Dr Otto Kamm, Alfred Payrleitner, Ekkehard Boesche, Alfred Winter, Bruno Modugno, Dr David Hodges, Janet Lamb, Erick Fernandes, Willem Beets, Dr John Picket, Dr Bob Evans, Barry Wookey, Ali Twaha, Lord and Lady Digby, Chantal Jourdan, Neil Sampson, Richard Hogg, Wes Jackson, Wendell Berry, Lawrence Woodward, Zikos and Sue Tassios, Vassilis Kalis, Prof. Hartmut Vogtmann, Prof. Engelhard Boehnke, Dr S.Heilenz, Prof. Mengel, Nikolaus Geiler, Martin Boehme, Prof. Peter Schuett, Prof. Juergen Kranz, Dr Hans Bibelriether, Dr Kurt Schefzik, Dr Bernward Geier, Dr Pius Stadelmann, Ulferd Dorka, Franz Obrubanski. There are many others who helped us for short periods of time and with information given over the phone; we are grateful to all of you.

Our special thanks to the governments of Kenya and Tanzania, to the National Museum in Athens, to Oxfam, Greenpeace and to the Liebig Museum in Giessen, West Germany.

Last but not least our thanks to our producer, Brian Turvey, and to the BBC for making the series – and this book – a reality. Special thanks to our editors, Valerie Buckingham at BBC Publications and Wolf Kugler at Fischer Verlag for working with us so constructively.

Herbert Girardet, June 1985

Introduction
- John Seymour -

When men and women first appeared on Earth they were part of the natural world about them and took their places among the other living creatures. They collected wild foods, hunted wild animals and they were hunted themselves. Gradually, over thousands of years, they developed the skills, tools and devices which gave them, as God, according to the Book of Genesis, had promised to the first man, 'dominion over the fish of the sea, and over the fowl of the air, and over every living thing that moveth over the face of the Earth'.

Now whatever one believes, one has to have an opinion on whether man is a part of nature or not. If he is, and in an age that has seen human genes successfully spliced with those of mice it is difficult to believe that he is not, then surely he must be very careful not to distort too grossly the biosphere of which he is simply a component. If he is *not* a part of nature, but divinely appointed to have dominion over it, then surely he should exercise humanity and restraint in his dealings with what he considers to be lesser creatures or lower forms of life.

It is frequently claimed that humans have the power to destroy every living thing on our planet. The characteristics of some viruses, algae and bacteria might put this in doubt, and the habitats I have seen of some creatures, for example cockroaches living 1000 feet underground in an African copper mine, might protect them from radiation, but we do undoubtedly have the ability to control or even destroy any of the higher multicellular forms of life, including ourselves. So far in our history we have never stayed our hand whenever it has suited us to destroy other forms of life. Whenever any fellow living creature has obstructed our profits, our pleasure, or our convenience we have always destroyed it without any compunction whatever.

Now the purpose of this book, and the television series which accompanies it, is to decide not what is ethical about mankind's treatment of other forms of life but whether, as an increasing number of people are beginning to believe, mankind's present exploitation of his planet is unsustainable. Can we continue to live as we are living, and work as we are working, for more than a limited number of generations?

It is incontrovertible that we are extinguishing more species of plants and animals every year that goes by. It has been suggested that the name for man should be altered to *Homo extinctor*. In the current destruction of particular habitats, notably the tropical rain forests all around the world, we are wiping out species that are not yet known to science and will never now be known.

It is possible to argue that this does not matter to us at all. We have safely domesticated the plants and animals that we require for our comfort and survival and, in environments controlled by ourselves, we can certainly retain such wild species as we feel we need for 'sport' (killing them for pleasure) so if the other species disappear the only consequence is that there is more room for us. Some might think this is sad but inevitable in the cause of what we call progress.

But humans have now spread nearly everywhere on this planet and their activities are causing more than the extinction of just a few species. Every acre of land that can be is either built on, or arably farmed, or grazed by our domestic animals; and both graziers and cultivators will not rest until they have destroyed every living thing on their holdings which they feel interferes with their profits.

Since farming began this attitude has always existed. What has changed is that previously graziers and cultivators did not have the technology to do it but now they do. Only in the last fifty years have we had the chemical and mechanical means to destroy effectively other forms of life on our planet. The other author of this book, Herbert Girardet, will develop later his concept of the amplified man: the creature whose own puny powers have been enormously amplified by his sudden access to vast stores of fossil energy.

It has been observed that power corrupts the wielder, and *Homo sapiens,* as man so immodestly calls himself, has almost absolute power and has ceased to exercise any restraint. The ancient customary and religious restrains have long since gone.

But we have not, in this book, concerned ourselves with ethics or with morals but with expediency. Is mankind, in pursuing his present courses, acting in his own best interests? Is our present course sustainable – or will it lead to disaster: disaster not only for other forms of life but for ourselves?

To answer these questions, Herbert Girardet and I have travelled many tens of thousands of miles, and visited many countries in four continents. The result is this book.

Humans and their Impact
——— John Seymour ———

My life began near the beginning of this century and I have seen dramatic changes in our relationship with the land. Anyone who has any interest in the land must view the immediate future with misgivings. To explain why this is so I find I can do no better than to recount a few of the salient experiences of my own lifetime.

I began my involvement with the land as a pupil on a 100-acre (forty-hectare) farm on very good, heavy land in mid-Essex. It was all arable except for a five-acre (two-hectare) paddock on which ran the horses during their short periods of rest. There were five Suffolk Punches, four of which worked in pairs ploughing or cultivating, or drawing the huge Essex harvest wagons, and the fifth was used for the odd one-horse jobs about the place. There was no tractor.

A quarter of the farm was always down to a one-year grass-and-clover ley. Grass seed was cheap in those days. After one crop of hay had been harvested, the ley would be ploughed up and its mass of roots and herbage returned to enrich the soil. The second quarter of the farm was sown to mangolds which were used for feeding the fattening bullocks; the third quarter was either down to spring barley or oats, or to vegetable crops grown especially for seed production. The remaining quarter was down to wheat which never threshed out at less than two tons to the acre.

Five men apart from me worked full time on the farm and we all worked long hours and extremely hard. We were all very fit. The farmer supervised us and attended to the management side of the business: I never saw him do any manual labour. In the autumn, 100 store (which means not fat) bullocks were bought and these were fed in covered yards until the spring when they were sold off fat. It was my job to feed them on chopped mangolds, chopped oats, straw, hay and a little linseed cake – the only product bought in to the farm. I also had to litter them with fresh straw so that they always lay clean and dry. They were contented animals and had the unmistakable bloom of health on them. When they went away in the spring they left behind them an almost unimaginable mountain of 'muck' as we called it, or farmyard manure, and all this had to be hauled out by the patient horses and spread upon the land. The farmer claimed that he did not make much money fattening the cattle: it was the muck he valued. It was this muck, together with that of six sows and their progeny, the horses, and about 100 hens, that made the farm so fertile. No artificial fertiliser was ever used and pesticides had not yet been invented. Because of the constant rotation of crops, and the innate fertility of the soil, disease in either plants or animals was practically absent. This old-fashioned farmer would talk about the 'heart' of his soil: above all he wanted it to be in 'good heart'.

Of course, such farming would be economically impossible now because of the high cost of labour. The land of the farm has no doubt been amalgamated into a much larger 'unit', white straw monoculture (wheat and barley) is practised, no four-footed animals are kept at all and any fertility comes from artificial fertiliser. As for 'heart' – if the artificial fertiliser supply was cut off for any reason the land would grow almost nothing at all and that response would be immediate, in the first year's cropping.

I mention my experiences on this farm for two reasons. The first is that it seems to me now quite remarkable that so much high quality produce was raised and exported from the farm with practically no outside input whatever. It was the land that was producing the output of the farm. The few tons per year of linseed cake that were fed to the cattle were about the only organic input – plus, of course, the 100 thin bullocks which went away, six months later, as 100 fat bullocks. The second reason is that I am quite sure that if that regime had continued indefinitely the soil would have retained or even improved its fertility. It would have lasted for as long as the world survived, or at least until some major geological upheaval, or drastic climatic change, altered the situation. I am quite certain that such a statement cannot be made about Essex land farmed by the methods in use today.

After three years at agricultural college I went to Africa. My first six months were spent on an enormous sheep farm in the Karroo – that vast plain on which trees and, generally, grass are absent, and the only flora consists of small edible shrubs scattered about the velt with bare ground in between. My employer was an enlightened man and very much concerned with preventing something of which I had never even heard before: erosion. He took me to see some examples of this on a neighbour's farm. I remember we rode our horses down into what was called a *donga*, an erosion gulley, and our heads were well below the rim of the gulley. For the first time I realised that soil is an expendable substance – it is fragile – it can disappear. I remember feeling a sense of shock at seeing the *dongas*. The stable world, the very ground beneath our feet, could be destroyed by greed and ignorance.

I did not realise then what I realise now, and that is we are creatures of the soil. A man is as much a soil organism as an earthworm is. If we include the plankton of the oceans with the soil of the land (as we should) then every single thing of which we are made derives from the soil. If scientists do succeed in producing some edible substance from oil or natural gas we will still be creatures of the soil, just as much, for both oil and natural gas are products of soils of long ago. Man has not yet learned to photo-synthesise and there does not seem to be the slightest prospect that he ever will. And to see the soil melting away under one's feet is a daunting thing.

My employer in the Karroo was, as I have said, an enlightened man. He farmed in a way which prevented much erosion on his soil. His sheep were kept in fenced paddocks (albeit each paddock was several miles across), and they watered at troughs supplied from deep boreholes – the water being pumped up by wind-pumps. Some erosion was caused by the sheep walking to and from the waterholes but this was minimised by increasing the number of boreholes. If ever there were any sign of a *donga* appearing on his land, the farmer quickly stopped its progress by fencing it off and filling it in.

My next experience of working on the land was in the northern part of what was then called South West Africa and is now named Namibia. There I managed a sheep farm on the edge of the desert. The farm was not even ring-fenced and so was contiguous with the rest of Africa. The veld, or landscape, that surrounded the farm was quite undamaged. The bush was intact. There was good grass between the little trees of the bushveld and the trees themselves were intact. Young trees and seedlings were able to grow to maturity because there were not too many grazing animals to destroy them. There were grazing animals, in plenty: zebra, gemsbok, kudu, springbok and many other species of antelope, but there were also beasts of prey to feed on them and keep their numbers within bounds. There was a balance. Over many thousands of years all the creatures of the soil – both animal and vegetable – had achieved a perfect balance. There were humans as well as other animals and these were beasts of prey, too, although they included wild plants as well as animals in their diet. They were men living in the Old Stone Age, who we called bushmen. Their small bands ranged over large areas of territory and I do not believe they made any impact that was discernible on either the soil under their bare feet or on the other creatures that lived on it.

I saw a good deal of these people for one of my shepherds was a bushman and he spoke the lingua franca of the country, Afrikaans. I spent much time hunting with this man, and also visiting his friends and rela-tions among the wild bushmen. The latter were hard to meet without an intermediary because they would simply vanish into the bush at the approach of a white man.

It seemed to me that the bushpeople swam in their environment as fish swim in water. As my shepherd, whom I called Joseph, moved through the bush he would not break off a single twig or disturb anything at all. He could track an animal over the hardest of terrain where I could see no trace of the animal's footprints and he would do this at top walking speed. I would sometimes ask him where the spore was and he would point incredulously at a fallen leaf that had been oh so slightly pushed into the soil or a pebble that had been turned over. He was incredulous because he found it hard to believe that anybody could be so blind as not to see these things. A white man, or a member of the pastoral or agricultural races of Africa, would have died very quickly in that environment. If Joseph was hungry he would dig out a corm or bulb from the ground with the sharp end of a gemsbok horn; if he was thirsty, he would find an insignificant little creeper on the ground, dig down underneath it to expose a spongy soft object from which he (and I) would suck some rather nasty-tasting water. I used to think that although it was nasty, it would have been nastier if it had not existed, because thirst comes quick and violent in that desert atmosphere. When we killed a gemsbok, as we often did, we would slake our thirst from the half-gallon or so of bitter water contained in the animal's paunch.

What I learned from Joseph, and my other Paleolithic friends, was to have a great respect for nature. They treated the other animals of the soil community as their equals. Lions once raided the farm and killed several of my employer's donkeys. Joseph and I spent the night on the ground beside one of the dead donkeys, knowing that the lions would revisit their prey. When they returned I shot one. Two others ran off and I wanted to follow them and shoot them too. We tracked them until they left our area and then Joseph refused to go any further. 'Why not?' I asked. 'We have punished the lions and they have gone away,' said Joseph. 'It would be wrong to punish them further.' I saw here a complete contradiction between the stone-age man's position and my own. For me, if it was good to shoot one lion it must be better to shoot three. To Joseph, what was needed was a *modus vivendi* – an honourable peace. The lions were as much a part of the community as we were and had an equal right to live. Yet we also had a right to stop them from killing our donkeys, just as we had a right to kill a buck for meat – but only when we needed one. Killing for killing's sake was inconceivable to Joseph. Joseph respected the animal he killed out of necessity. Joseph's relatives danced at night, by the firelight, closely imitating the animals they hunted by day. Joseph's ancestors had depicted these same animals, often being hunted, on the walls of caves all over Africa. Modern bushmen no longer do this because they have been driven into country where there are no caves – no rocks.

Very different from the territories of the wild bushmen was the country that had been given over to the settlement of white people. The farm over which I ruled, being at the frontier of white settlement, and not fenced at all, had no real boundaries. The size of the grazing area I had at my disposal was limited only by the distance that a sheep could walk – and return again – in a day. The sheep slept in kraals, or fenced compounds, during the night, to be safe from wild animals, and were protected during the day by African shepherds who were constantly with them. Each night the sheep returned to the homestead to drink water and to be penned in.

The veld near the homestead was therefore subjected to very heavy pressure. The sharp hooves of 2000 sheep trod out the grasses, the many nibbling mouths snapped up any seedling tree which reared its head above the ground. As the old trees died there would be nothing to replace them. On the old-established farms, settled by German settlers in the nineteenth century, the long-term effects of this could be seen. Here the veld was tramped out as we called it, for miles around the farmstead. The sheep had further and further to go to get to the grazing and the further they walked the more they destroyed the grazing on which they depended. It was quite obvious here that the grazing, on which sheep could live and farmers grow rich, was a wasting asset just as the coal measures are or the oil-fields. You could go on using it until it was finished. It would not replace itself again.

I compared this state of affairs to two things. One was the unsettled wilderness where the bushmen roamed. That could go on, I realised, until the end of the planet without deteriorating. The great variety of plants and animals that inhabited it probably represented the highest biological output of which that land, in that climate, was capable. The 'white-owned' farms were obviously a wasting asset. The effect of the new owners was extractive and destructive.

I compared it, too, to the farm on which I had worked in Essex. *That* farm, as I have stated before, could have gone on until the end of the world without deteriorating, for there, as in the Namibia and Kalahari bush countries, was a sustainable land-use regime.

Later, I wandered to many countries, both in Africa and Asia, and except for service in the army during the Second World War was nearly always, in some way or other, connected with the land.

I found many obviously sustainable methods of agriculture. In central Africa and in the northern jungles of Sri Lanka I saw what the inhabitants call *chena* cultivation. The farmer moves into the forest, cuts and burns the trees, sows his crop seeds among the ashes and for perhaps four or five years enjoys good crops. Then, exhaustion of the soil, combined with the invasion of arable weeds, makes farming unproductive so the farmer abandons that land and clears more forest elsewhere.

This may sound grossly destructive but, provided the cultivators are not too numerous, it is not. For the forest rapidly recolonises the clearings, grows for a few decades and then, by the time the *chena* cultivators return, the fertility has been built up again. There is another round of slashing and burning and four or five years more cropping done.

Alas, when people become over-numerous this system breaks down. First, the cultivators stay too long in a clearing, thus impoverishing the soil to vanishing point preventing the forest from returning, and second, they return to a formerly cropped area too soon. The jungle has not had time to recover itself. This inevitably leads to complete soil degradation, erosion and ultimate desertification. And it is happening, now, in all too many areas.

I saw also successful and well-managed paddy cultivation in Sri Lanka and elsewhere and that, it seemed to me, is indefinitely sustainable. The soil cannot be eroded by water, because it is under water and the water is stationary. It cannot be eroded by wind for the same reason. Provided the manure from animals and from humans is returned to the paddy-fields, fertility is kept up.

I saw fine farming on the flat plains of the Punjab in northern India, where the methods of husbandry seemed to me just as sustainable as were the methods of the farm in Essex. Again, the law of return was observed. What was taken from the soil was put back, in a different form, into it. The humus content of the soil was strictly conserved and renewed. I saw fine arable farming, at 8000 feet above sea-level, by the Christian farmers in Ethiopia. The same land had been farmed for 1000 or more years yet there still seemed to be no diminution in soil fertility, or serious soil erosion. I also saw, in the low-lying arid country surrounding the Ethiopian high-lands, a nomadic pastoral way of life that had survived for millennia. Again, provided the human and grazing animal populations remained fairly stable, this could have gone on for ever. Alas the populations did not remain stable and the results can be seen in the recurring famines in Ethiopia and other parts of the Sahel. The population is now being adjusted drastically, as populations often are, by a limitation of the food supply, which is sad; but, sadder still, is the fact that due to overgrazing the soil had gone. The Sahara Desert has shaken its shoulders and enlarged itself.

I came back to Europe after my wanderings, thinking that at least I was returning to a continent which, because of its moderate climate and good tradition of farming, was stable and secure. I have spent the last few years gathering material for this book and the films of which it is a result, mostly in Europe, and I now have very serious misgivings indeed. Man, as a species, is rapidly destroying the community of which he is a part: the living community of the soil.

The Long Beginning
—— Herbert Girardet ——

Who can deny that in the final analysis humans are creatures of the soil? We are reminded of this with each meal we eat, every day. And yet in this modern world most of us are utterly removed from a soil-related way of life. The industrial societies of the northern hemisphere pride themselves on having reached a point where only a small proportion of the population is still actively involved in food production. The fewer people working the land, the more advanced we are considered to be. After all, who wants to be a slave to the seasons, to the whims of nature? For thousands of years – at least since the very first days of agriculture – we have been involved in a struggle to dominate nature, spending our lives in an effort to overcome biological limitations. Now, most of us seem to want to forget that farming or foraging for food was the 'normal' way of life for many people until very recently.

Our 'contemporary ancestors', the last surviving hunter-gatherers such as the bushmen, may be physically almost identical to us but that is where the similarity ends. In ecological terms, a bushman or a pygmy – or even a subsistence farmer – is a totally different creature from a late-twentieth-century technological man. The pygmy is a creature of the tropical jungle; we have become creatures of the concrete jungle. The pygmy acknowledges his thankfulness to the forest which is his home and his god in the daily conduct of his life. Modern technological man has utterly forgotten that he owes his life to the continuity and fruitfulness of nature; he has come to worship himself and his own products.

The pygmy or the bushman is what he appears to be: an essentially biological being. He takes from the bounty of nature what he needs for his daily living, no less and no more. He participates in the cycle of growth and decay as part of nature. His metabolism and his environmental impact are similar to those of the animals of a like body size with whom he shares his habitat. His possessions are limited to what he can carry from one camp to the next. The hand tools which he uses enable him to harvest plant and animal foods, to ensure the continuity of his tribe while allowing it to maintain the integrity of its host environment.

Modern technological man, the 'amplified man', is a new kind of creature. He is half-man, half-machine. He has human arms and hands for pressing buttons and manipulating knives and forks, for operating gear levers and steering wheels. His feet have been replaced by wheels, his

19

muscles by petrol engines and electric motors. His body is flesh and metal, in his veins blood is mixed with lubricant oil. His breath consists of sulphur dioxide, nitrogen oxide and the many gases that are emitted from smoke-stacks and exhaust-pipes. His tentacles of electric wires and telephone lines stretch around the world and even into space. Computers have been grafted on to the circuity of his nervous system. And yet his brain, heart and sexual organs are still riddled with the remnants of 'primitive' human emotions. His metabolism is made up of all the biological, technical and chemical processes by which his industrially-based life-style is maintained.

This transformation of man is the product of a history that started as recently as 12,000 years ago when agriculture and then mining and metal-working were first taken up. That is a very short timespan in the life of mankind. In East Africa, anthropologists have found fossilised remains of our early ancestors that date back around two and a half million years. The evidence suggests that their way of life was very similar to that of surviving hunter-gatherer tribes. Some – like the bushmen of today – were inhabitants of the open forests that still exist, the savannah woodlands in which acacia trees co-exist with grasses and shrubs, a food source for innumerable grazing and foraging animals. Others – like the pygmies who still roam the Ituri Forest in Zaïre, central Africa – lived in dense tropical rain forests in a world of cool shade and flickering sunlight. Paleoanthropologists, like the Leakey family who mostly work in Kenya, have established that some of our earliest hunter-gatherer ancestors were physically very advanced, walking upright as we do. Their hands gripped in the same way as ours and their brains were almost as large as ours.

In recent years there has been a dramatic change in our understanding of hunter-gatherer societies, ancient and contemporary. Anthropologists working in Africa, Asia, America and Australia have been able to prove that hunter-gatherers still remaining today do not lead 'short, nasty and brutish' lives as was widely assumed until quite recently. Quite the contrary. Hunter-gatherers – after careful study – have been described as the 'original affluent society'. They rarely go hungry: even in 'marginal environments' such as the Kalahari Desert in south-west Africa they manage to feed themselves very adequately without too much effort. They usually live in groups of thirty or forty people in temporary camps where they stay until the food supply in the surrounding countryside is used up. Men do most of the hunting but, together with women and children, will also join in with the gathering of plant foods and small animals like mussels and snails. The food tends to be more varied than that of 'primitive' farming groups. Women usually build the communal huts or shelters out of branches and leaves found in the vicinity.

Hunter-gatherers are not rich if the accumulation of possessions is the measure. They tend to own no more than they can carry around with them from one camp to the next. But their mobile lives are rich in variety, much more so than those of sedentary societies. There is plenty of adventure and also a great deal of leisure since usually no more than two to four hours per day are spent on procuring food.

Hunter-gatherers are neither stupid nor mentally underdeveloped as has so often been suggested. Their brains are as big and complex as our own and who are we to suggest that they do not make use of them to good effect because they have not invented the space-shuttle or the atom bomb? They must know and understand the environment they inhabit, its rock formations and watercourses, its caves, springs, plants and animals, if they want to survive in it. Their survival through enormous spans of time is the best possible proof of their great range of knowledge of a largely unmodified environment. Pygmies in the tropical forests of Africa have names for every type of plant that grows in these surroundings. This is also true of the 'primitive' Indian tribes who live in the jungles of South America which are botanically among the most diverse regions anywhere in this planet. The children of forest-dwelling Amerindian tribes – who live by hunting, gathering and gardening – will know up to 1000 plant names by the time they cease to depend on their elders. Claude Lévi-Strauss, in *The Savage Mind*, quotes a biologist who lived among the forest-dwelling pygmies of the mountain ranges of the Philippines:

Another characteristic of Negrito life, a characteristic which strikingly demarcates them from the surrounding Christian lowlanders, is their inexhaustible knowledge of the plant and animal kingdom. This lore includes not only a specific recognition of a phenomenal number of plants, birds, animals and insects, but also includes a knowledge of the habitats and behaviour of each . . . The Negrito is an intrinsic part of his environment and, what is more important, continually studies his surroundings. Many times I have seen a Negrito who, when not being certain of the identification of a particular plant, will taste the fruit, smell the leaves, break and examine the stem, comment upon its habitat and, only after all of this, pronounce whether he did or did not know the plant.

Lévi-Strauss, who himself carried out field-work among Indians in the Amazon forest, stresses that the forest people are not only interested in the plants that are of direct use to them but consider it important to develop systems of plant classification. He points out that the knowledge of the ordinary tribespeople is far exceeded by the biological knowledge of medicine men and women who use plant remedies constantly in their daily practice.

Knowledge comes through exposure to a given environment, through theoretical instruction and through experience. Who are we to suggest that this applies only to so-called advanced civilisations? Why is it that the biggest and richest chemical companies in the world send scientists into the tropical forests and ask nearly naked 'savages' to point out the medicinal uses of vines and mushrooms, flowers and leaves of bushes and trees? The huge variety of botanical treasures that are stored in tropical forests, which are now being destroyed and degraded at the rate of about sixty acres, or twenty-five hectares, per minute, are categorised and labelled by their inhabitants who have been called the 'librarians of the forests'. They themselves are now under threat of extinction as the pressure of the civilisation of the 'amplified man' increases.

Before we finally remove tropical forests and forest people from the face of the Earth, we should remember that we owe to them much of the knowledge of food plants that we have come to rely on for our agricultural and plantation crops. Many of our most important food crops and medicinal plants originate from tropical forests, notably the jungles of South America, and were used for millennia by their inhabitants. In a recent article, Dr Conrad Gorinski stated:

The Amerindian supplied the potato and thereby helped feed the industrial revolution in Europe as well as providing cassava, now a staple food of Africa and tropical Asia. Peanuts and cocoa are also of Amerindian origin. Cocaine derivatives are used as local anaesthetics and curare helped give surgery the status it holds today. Quinine helped launch the field of chemotherapy, and rubber, which also originates in the Amazon Forest, made much of today's technology possible. For sophisticated palates, the Amerindian provided tomatoes, cashews, Brazil nuts and avocados – to name a few.

Many other crops from tropical forests, known and used by forest dwellers for food and medicines for time immemorial, are at present being investigated for their wider potential by scientists of many countries and disciplines.

When concerning ourselves with forest people and their knowledge, we should not only consider their usefulness to us. It is also intriguing to realise that forest dwellers such as the pygmies or some of the North American Indian tribes have elaborate ceremonies which emphasise the 'goodness' of their forest home and their thankfulness for being able to inhabit such a place. The anthropologist Colin Turnbull spent many years in the company of the BaMbuti pygmies of the Ituri Forest in Zaïre. In his book *The Forest People* he describes in great detail their belief in the wonderful goodness of their forest. One of them told him:

The forest is like a father and mother to us, and like a father or mother it gives us everything we need – food, clothing, shelter, warmth . . . and affection. Normally everything goes well, because the forest is good to its children, but when things go wrong there must be a reason.

Another one added:

At night when we are asleep sometimes things go wrong, because we are not awake to stop them from going wrong. Army ants invade the camp, leopards may come in and steal a dog or even a child. If we were awake these things would not happen. So when something big goes wrong, like illness or bad hunting or even death, it must be because the forest is sleeping and not looking after its children. So what do we do? We wake it up. We wake it up by singing to it, and we do this because we want it to awaken happy. Then everything will be well and good again. And when our world is going well then also we sing to the forest because we want it to share our happiness.

The ancient songs that the pygmies sing to their forest were recorded by Colin Turnbull. They are of great melodic complexity and haunting beauty. Nobody knows for how many generations they have been taught by parents to their children and grandparents to their grandchildren. There are strong indications that similar songs were already sung when the Egyptians first encountered the pygmies – around 2500 BC – during the rule of the Pharaoh Nefrikare in the course of an expedition up the Nile to find the source of that great river in the mountains and forests of Central Africa. Colin Turnbull records the evidence:

In the tomb of the Pharaoh Nekrikare is preserved the report of his commander, Herkouf, who entered a great forest to the west of the Mountains of the Moon and discovered there a people in the trees, a tiny people who sing and dance to their god, a dance such as had never been seen before.

This god of the pygmies, as we have seen, is their own forest which supplies them with all they need to live and whose continued existence is in their vital interest. Since the moisture-laden forest also ensured the continued flow of the Nile, and with it the continuity of their own great civilisation, the ancient Egyptians depicted the pygmies in some of their wall paintings with great reverence as the wise people of the forest. Their hunter-gatherer way of life assured the integrity of the forest. Today, as the jungles of central Africa – like the rain forests throughout the tropics – are shrinking rapidly under the pressure of 'progress', it is not only forest people like the pygmies who are under threat, but also the life-giving flow of great rivers such as the Nile on which many millions of farmers and cattle herders depend for their survival.

A detailed understanding of nomadic hunter-gatherer societies is important for us not only because they represent the roots of humanity, geneti-
cally as well as culturally, but because they have had a 'bad press' for so
long, being considered as examples of the worst savagery and undesir-
able primitiveness. Being settled is thought to be one of the preconditions
of being 'civilised', that is, not to be beckoned too much by the 'call of
nature'. Understanding nature is the be-all and end-all of the hunting-
gathering way of life, so much so that the local territory is imprinted on the
individual's mind like a three-dimensional map of the world, the tribe's
local world. This point was made very clearly by Gary Snider in his 1982
Schumacher Lecture in Bristol in which he described his travels in the
company of some Australian Aborigines:

I was travelling by truck over a dirt track west of Alice Springs in the company of a Pintubi elder named Jimmy Tjungurrayi. As we rolled along the dusty road, sitting in the back of a pick-up truck, he began to speak very rapidly to me. He was talking about a mountain over there, telling me the story about some wallabies that came to that mountain in the dreamtime and got into some kind of mischief there with some lizard girls. He had hardly finished that when he started into another story about another hill over there and another story over there. I couldn't keep up. I realised after about half an hour of this that these were tales to be told while *walking*, and that I was experiencing a speeded-up version of what might be leisurely said over several days of foot-travel . . . So remember a time when you journeyed on foot over hundres of miles, walking fast and often travelling at night, walking night-long and napping in the acacia shade during the day, and these stories were told to you as you went. In your travels with an older person you were given a map you could memorise, full of lore and song and also practical information. Off by yourself you could sing those songs to bring yourself back. And you could travel to a place that you'd never been to, carrying only songs that you had learned . . . This is the way to transmit information about a vast terrain which is obviously very effective, and doesn't require writing.

In desert-like conditions such as those that are found in some parts of Australia or the Kalahari region of south-west Africa, the hunter-gatherer way of life requires great mobility. Groups of thirty to forty people will move around together. These are usually family and kinship groups that consist of three generations. They are capable of meeting their own needs for food, water and firewood. Land is not owned by individuals or even the group. A tribe is usually considered to have rights over a certain area or land and the food resources – plant and animal – contained within it. Even in harsh desert conditions, the lives of hunter-gatherers are not ruled by necessity alone. For the Australian Aborigines, some areas of land such as special rock formations or groves of trees are sacred, to be considered with great awe and to be visited only for important ceremonies. Today, the Aboriginal tribes go to great lengths to prevent mining companies from desecrating their holy places with excavating machinery. But our appetite for uranium, copper and iron ore is such that the mining companies usually succeed in the end.

Not all ancient hunter-gatherers would have lived in a paradise on earth but some tribes would have had a very pleasant existence indeed in an environment of great fertility. The Garden of Eden, or Dilmun as it is called in ancient Sumerian texts, can be considered to have been a real place, the 'home ground' of the likes of Adam and Eve for many, many millennia. The story of the expulsion from paradise in the Old Testament could be described as the introduction to the history of a tribe that had only recently made the transition from a food gathering existence to an agricultural/ herding way of life. Adam's tribe probably lived in the open oak wood-lands in the foothills of the mountain ranges of the Near East. According to Genesis, God had given man for his food 'all the seed-bearing plants on

24

earth and all the trees that bear fruit containing seed'. The lower mountain forests of the Near East contained oaks, pistachios, almond trees and a variety of fruit trees as well as primitive forms of wheat and barley. Lentils, peas, rape seed and various other food plants also originated there. There is no doubt that all of these would have been used by food gathering tribes long before the process of domestication began. Wild sheep and goats lived in the mountain ranges and hills of what is now Iraq, Iran, Turkey, Syria, Lebanon and Israel. Domestication of food, both plants and animals, seems to have been a very gradual process, starting tens of thousands of years ago. The earliest settlements that have been excavated date back approximately 10,000 years. In some of them only traces of vegetable foods have been found, but in most there are traces of the bones of sheep and goats as well.

The village of Jarmo in the foothills of the Zagros Mountains in northern Iraq is one of the best-known examples of an early human settlement. It was excavated in 1950 by Robert Braidwood and other members of his 'Prehistoric Project', a research team from the University of Chicago investigating the earliest agricultural settlements. It is located on top of a bluff at an elevation of 8750 yards, or 8000 metres, a height which is quite typical of early farming settlements in the Near East. Jarmo lies in the hilly country east of the modern town of Kirkuk. The remains of this prehistoric community cover an area of about two and a half acres, or one and a half hectares, and are preserved to a depth of approximately seven yards. As many as twelve layers of buildings have been identified, representing a community of between 150 and 200 people. There were successive settlements on this site for several hundred years. The quite substantial buildings consisting of several rooms were made of compacted mud and were rectilinear. Many of them had small courtyards. The village site has been dated to 6750 BC.

The people of Jarmo grew barley, einkorn and emmer wheat which are indigenous to this area. They also grew legumes such as peas, lentils and chick-peas. Tools used for collecting and processing both wild and home-grown food plants were found in considerable numbers during the excavations. They included flint sickles, mortars, handstones and ovens which were probably used for parching grain. Stone bowls were unearthed which were almost certainly for food storage. Objects made of clay were also found though they seem to have been made mostly for decorative purposes. Stone axes and adzes were in use, too.

The archaeological evidence suggests that agriculture started in a variety of places more or less simultaneously some time after the end of the last Ice Age, around 9000 BC. Traces of early agriculture have been found in the Near East, Egypt, Greece, China, Thailand and South America. Nobody has been able to explain this phenomenon with certainty. Some researchers claim that agriculture first started in river valleys, where stable settlements developed on the basis of fishing. Others are convinced that hillsides with regular rainfall were the most suitable environment for early farming. Inevitably these settlements led to population growth and environmental change.

The greatest impact of early agricultural settlement was on hillside forests. The open oak forests of the 'fertile crescent' were the watersheds for the two great rivers, the Euphrates and the Tigris. Villages sprang up throughout these hilly regions. Trees were felled to make room for fields for growing grains and legumes. Animals, goats and sheep were herded through the forests in search of grazing. House construction needed timber and the development of pottery and metal-smelting required firewood and charcoal. Populations of humans and domestic animals grew to the detriment of the forests – and therefore of the wild animals.

Many villages sprang up in the upland regions above the Mesopotamian plain between 7000 and 4000 BC. There is no doubt that the rapid development of agriculturally based civilisation had a very considerable effect on the hillsides and mountain forests of the Near East. Agriculture implies the annexation of a given area of land for the exclusive use of man and the removal of trees and shrubs that might interfere with cultivation. If this is done over a wide area on sloping ground destabilisation of the soil is inevitable. The only way to avoid this is to terrace the land but no traces of terraces have been found in the vicinity of these early settlements. The impact of agriculture on forests and on the hillside soil was further increased by the growing populations of domesticated animals such as sheep and goats which prevented the regrowth of trees on the hillsides through their grazing. There is little doubt that the bare landscape we see today all over the hilly regions of the Near East is the direct result of deforestation for agricultural and pastoral development, dating back to the beginnings of plant and animal domestication.

The impact of growing farming populations on hillside and mountain forests can be observed today in places such as Nepal. Here, peasants are deforesting the slopes of the Himalayas at a breathtaking rate, as a consequence of the growing demand for agricultural land, fuelwood and timber. The remaining forests are being destroyed rapidly because pressure from grazing animals prevents the regrowth of trees. It is interesting to note that Nepalese peasants have no notion of setting up tree nurseries and planting trees to ensure the continuity of the forests whose great value they readily acknowledge. It is likely that early farmers in the Near East were similarly unfamiliar with the idea of replacing trees that had been cut down.

It seems that villages like Jarmo could be maintained for only a few hundred years, succumbing in the end to erosion and loss of soil fertility. Today, the area around Jarmo is wasteland, like so much of the hilly country above the Mesopotamian basin. Not all of this devastation was caused by early agricultural experiments: overgrazing by sheep and goats continues to this very day.

The importance of maintaining tree cover on sloping land cannot be stressed too strongly, since this is a matter which is insufficiently understood even today. The roots of trees on mountains or hillsides go deep down into the cracks of the rocks, stabilising the soil. The hair-roots just below the soil surface bind the particles together, ensuring that the soil will not be washed down the mountainside during rainstorms. It has been repeatedly shown in studies conducted in many countries that slopes that have been denuded of trees suffer from greatly increased erosion problems. It is also a well-established fact that forest soil acts like a sponge, soaking up the rainwater and releasing it slowly down the mountain slopes into the rivers and valleys below, giving a steady supply of water. Forests also produce a higher level of atmospheric moisture than denuded land, and on hills and mountains they will tend to stimulate the release of rain from the clouds.

The removal of forests from the hillsides of the Near East which was set in motion by early agricultural communities was greatly accelerated when cities started to spring up on the plains between the Euphrates and Tigris. Mesopotamia, 'the land between the rivers', was virtually treeless and the cities that were built from around 5000 BC onwards would have had to import all the timber they used from the Zagros Mountains, the Armenian Highlands or wherever else it was available. There is little doubt that deforestation of the headwaters of the two rivers would have been convenient since it would have been easiest to float logs down from there towards the new cities. Eridu, Uruk, Ur and all the other early cities were built on the shores of one of the two great rivers.

The excavation of the Sumerian city of Ur by Sir Leonard Woolley in the 1920s, 1930s and 1940s was one of the major events in the history of archaeology. There are, of course, references to early cities such as Ur, Eridu, Jericho, Babylon and Nineveh in the Bible, but until excavations started no one had suspected that the remains of these places would ever be found.

The excavations at Ur and other Sumerian cities brought to light a rich variety of artefacts from which it is possible to reconstruct the life-style of their inhabitants. Ruth Whitehouse, in her book *The First Cities*, described it as follows:

We know that most of the population of a Sumerian city such as Ur was engaged in the production of food. Most of the people we passed in the streets would be farmers, market gardeners, herdsmen or fishermen and correspondingly many of the goods being transported in carts would have been food products. However, some of the farmers would have had other roles as well: carpenters, smiths, potters, stone-cutters, basket-makers, leatherworkers, wool-spinners, bakers and brewers are all recorded, as are merchants and what we might call the 'civil service' of the temple community – the priests and the scribes. But . . . probably everyone was first and foremost a practical farmer and at seed and harvest time every able-bodied man was no doubt on the land.

Agriculture laid the foundation of urban life and in Mesopotamia the soil and the water which made intensive farming possible was a gift of the mountains and the hills. From time immemorial the annual floods of the Euphrates and Tigris had brought some silt down from the mountains. These supplied the early farmers of the plains with fertiliser for their fields and with irrigation water. Reduction of the forest cover in the uplands because of land clearance and logging accelerated the transport of soil from the hills down into the river valleys and was thus in many ways beneficial to the Sumerian farmers. There are some indications that it was a process which was regarded favourably and even deliberately encouraged by the inhabitants of these early cities of the plains, who may have been unaware of the longer-term ecological consequences.

A similar process is taking place today: deliberate deforestation and hillside erosion by valley farmers has been reported from the Nochixtlan area of Mexico. The local Mixtec farmers have used this process for the last thousand years or so in order to produce deeper layers of topsoil on their fields and to increase the cultivation area in their valley. Judicious use of gully erosion enabled them to turn poor hilltop fields into rich farmland in the valley below. Soil and silt of an average depth of five yards was removed in this way from the surface area of the hills.

It is likely that something similar occurred in the Mesopotamian basin, but on a larger scale. When Sir Leonard Woolley excavated Ur he was astonished to find that after his workmen had dug through one yard of soil littered with potsherds and other human artefacts there was a three-yard layer of virgin soil and then below that, more traces of human activity. What could this be? Could it be . . . Noah's flood? This was a theory which was widely discussed at the time and which still has some followers today. (Noah – according to Genesis – was an ancestor of Abraham and was born in 'Ur of the Chaldees'.) After the inundation of Ur, which is thought to have occurred at around 2500 BC, the city was rebuilt three yards higher up and was an important centre of civilisation for many hundreds of years. There is little doubt that the three-yard mud deposit at Ur and those found at other Sumerian cities (though not necessarily from the same period) represent one of the earliest man-induced environmental disasters. The appetite of the Sumerians for timber, for silt and irrigation water had backfired.

The spectacular mud deposit at Ur is an indication of the newly-found power of man to affect his environment. Here, for the first time, we catch a glimpse of the 'amplified man'. The environmental impact of the Sumerians resulted from a combination of agricultural and urban development. The growth of their cities would have been impossible without a sophisticated system of food production. For this, four basic requirements had to be met. First, an assured supply of fertilisers. The Sumerians – like the Egyptians – did not opt for recycling their own wastes in order to ensure continued soil fertility, preferring to make use of silt washed down from the mountains. Second, a predictable flow of irrigation water for the crops. The distribution of irrigation water to the fields required complex engineering works and a well-developed administrative system. Third, adequate storage facilities for the harvested crops. There were storage places for grain (barley and wheat) in every Sumerian city, ensuring a regular supply of basic foodstuffs for the inhabitants as well as buffer stocks for times of scarcity. Finally, an agricultural planning policy. The Sumerians had an elaborate calendar which was used to plan sowing and harvesting and other important agricultural activities. They also had a 'farmer's almanac' which explains in great detail how people should cultivate a field:

When you are about to take hold of your field [for cultivation], keep a sharp eye on the opening of the dikes, ditches and mounds [so that] when you flood the field the water will not rise too high in it. When you have emptied it of water, watch the field's water-soaked ground that it stay virile ground for you. Let shod oxen trample it for you; after having its weeds ripped out [by them and] the field made level ground, dress it evenly with narrow axes . . . Let the pickaxe wielder eradicate the ox hooves for you [and] smooth them out . . . When you are about to plough your field, let your plough break the stubble for you . . . Keep your eye on the man who puts in the barley seed. Let him drop the grain uniformly two fingers deep . . . After the sprout has broken through [the surface of] the ground, say a prayer to the goddess Ninkilim [and] shoo away the flying birds. When the barley has filled the narrow bottom of the furrow, water the top seed . . . (From *The Sumarians* by Samuel Noah Kramer).

This farmer's almanac, which was inscribed on one of many thousands of clay tablets which were unearthed in Mesopotamia, goes on to explain how harvesting should be conducted and how the barley should be threshed and winnowed.

The early cities of the Near East were the first centres of wealth in human history. This wealth was based on careful planning and organisation of the productive system. In addition to the farmers, there were brick-makers, potters, carpenters, metal-workers and a great variety of other craftspeople. In the early stages of urban development they organised themselves on a fairly egalitarian basis. But it seems that growing wealth resulted in expressions of envy by neighbours and warfare between the cities. All of the early cities were protected by ring walls and by bowmen and heavily-armed soldiers. Warfare became a regular feature in the life of the people of Sumeria and of other urban centres in the Near East.

These city states not only had to protect themselves against attacks by other cities but also against nomadic tribes, for example the Gutians, who would come down from the hills or from distant regions to plunder the municipal stores. The nomadic tribes were greatly feared because they were able to use their mobility with deadly effect in surprise attacks. They were pastoralists who depended on herds of cattle, sheep and goats for their livelihood.

The Old Testament describes the movements of the pastoral Israelites in great detail and emphasises the conflict of interest that existed between the settled and the nomadic life-style. (The fatal clash between Abel, the herdsman, and Cain, the farmer, is a first example of this.) Genesis gives a fascinating account of the changing fortunes of the Israelites in their travels between the centres of civilisation in the Near East. Their herds were very vulnerable to drought conditions when they would run short of grazing and fodder. In times of drought and famine they would tend to seek refuge in well-watered places such as the valleys of the Euphrates or the Nile.

The story of the seven fat years and the seven lean years is an intriguing example of how environmental conditions can bring political change. Joseph, the son of Jacob, had been sold by his brothers to Egypt where he succeeded in becoming the Pharaoh's prime minister and right-hand man. He predicted that there would be seven years of famine. The Pharaoh decided that one-fifth of the harvest during the preceding fat years should be collected and stored as a buffer stock.

When the drought became severe, the Pharaoh was in a position to sell grain to the people of Egypt (and also to tribesmen such as the Israelites), in exchange for silver and gold they had accumulated. When precious metals had run out they had to sell their herds of cattle and sheep in exchange for wheat and barley. Finally, towards the end of the drought, the Egyptians were forced to sell their plots of land to the Pharaoh in exchange for grain. From then on, he and his successors were able to extract land rent from their subjects, thus greatly increasing the political and economic powers of the ruling dynasties.

This well-known story illustrates how people, in the course of the development of civilisations, came to depend increasingly on man-made systems of food production and distribution and on centralised power structures. Food came no longer from the untamed bounty of nature, as had been the case in days of hunting and gathering, but from the clever manipulation of the environment. Accurate forecasting of droughts and famines enabled rulers to increase their hold over their subjects. The god-like power which the masters of the early cities of Egypt and Mesopotamia acquired for themselves gave rise to bizarre delusions of grandeur, still visible today in the form of pyramids and ziggurats. The quest for immortality is a well-established feature of the psychology of the rulers throughout the ancient Near East. Their quest for power over people went hand in hand with the belief that they could dominate and control nature for their own purposes.

It is still not fully understood why the silt-based agriculture in the Nile valley has survived to this very day (or at least until the building of the Aswan dam in the 1960s), while agriculture in Mesopotamia, in the plain watered by the Euphrates and Tigris, collapsed more than 2000 years ago. It is clear, however, that the silt load of the twin rivers has always been much greater than that of the Nile, due to the virtually total denudation of the upland regions (the Armenian Highlands), from which they rise. As a consequence of the growing silt burden which the rivers carried with them, the shoreline of the Persian Gulf shifted 130 miles further south in the course of just a few thousand years. There is no doubt that the Sumerians considered the mountain silt as a curse as well as a blessing since it was for ever clogging up the irrigation canals. And the Euphrates acquired the habit of frequently changing its course in the flat plain, overflowing its banks and making a new bed for itself. This made life in the cities on its banks very difficult.

The silt load of the Nile, in contrast, was more moderate. Most of it came, as it still does today, from the Abyssinian Highlands – the headwaters of the Blue Nile – which were quite densely forested. Even forty years ago nearly half of these Highlands were still covered with forest. Since then, deforestation in Ethiopia has proceeded at a breathtaking rate, leading to both soil loss and drought conditions. Today, only about 4 per cent of the Abyssinian – or Ethiopian – Highlands are still forested, and much of the Highland soil seems destined to end up in the Aswan dam.

There is evidence that the decline of agriculture in Mesopotamia was not only due to silting problems and warfare. An additional factor, which is now being studied very carefully, is the build-up of salt in the soil. There are salt deposits among the rocks of the Armenian Highlands which have been washed down with the floodwaters. The slightly saline irrigation water used by the farmers of Mesopotamia left salt deposits in the soil when the hot summer sun dried out the fields. This seems to have led to a steady decline in soil fertility.

The Sumerians recognised the destructive effects of saline soil and combated them by limiting crops to one harvest per year (two were usual in the early period), and by growing crops which have a high tolerance of salt. The date palm has a particularly high tolerance approaching 2 per cent . . . while barley can tolerate nearly 1 per cent, wheat less than 0.5 per cent . . . In about 2400 BC, wheat formed 16 per cent of the total cereal crop (composed of wheat and barley together); by 2100 wheat had dropped to about 2 per cent and after 2000 wheat does not appear in the records at all. (Ruth Whitehouse, *The First Cities.*)

The salinisation of the soil of Sumer could be described as the first major chemical pollution disaster in the history of mankind. Salt is a 'natural chemical' but without intensive irrigation it would never have pickled the soil of ancient Mesopotamia.

Today, the twin problems of upland deforestation and salinisation of soil under irrigated agriculture are recurring throughout the tropics and sub-tropics. Salinisation will be discussed in more detail in a later chapter.

When Europeans first travelled extensively in Mesopotamia in the nineteenth century there were few traces to be found of the ancient cities, only mounds of sand and rubble in the desert landscape. Only the excavations by archaeologists revealed the magnificent remains of these ancient civilisations. The hubris and arrogance of the rulers of the Sumerian cities turned out, in the end, to have been self-defeating. Their magnificent cities and their elaborate irrigation works succumbed to mountain silt and desert dust.

The First Villages and the First Cities
—————————— John Seymour ——————————

It is true to say that on this planet now there are humans working at every stage of agricultural development. In a world that has electrostatically-operated, no-tillage herbicide-spraying machines, you may still find people using the digging stick. I have frequently watched bushpeople in Africa digging up edible corms with gemsbok horns and I am told that in Borneo there are people who fell trees with stone axes and dig with stone hoes. Until recently, the Bantu peoples of east, south and central Africa have been living very happily in the Iron Age.

The Balovale people, who live around the headwaters of the mighty Zambezi River, still smelt iron in their age-old way. The Barotse people (properly called baLozi or Lozi) who live in the Western Province of Zambia, use iron axes to chop down trees to make cultivable clearings in the forest and then use iron hoes to work the ground. When I lived in that country before the Second World War, the symbolic emblem of a man was a tiny ornamental axe and that of a woman was a similarly-sized hoe. The older hoe- and axe-heads in use were made of Lovale iron, the newer ones of European steel forged in exactly the same traditional shapes. These people, because of their contact with the Europeans, were gradually moving from the Iron Age to the Steel Age. Steel items, ox ploughs and European cloth were traded in exchange for such objects as hippo hides cut into strips for machine belting, ivory, buffalo hides and the skins of otters, leopards and other animals.

But I had the privilege of seeing the Barotse culture as it was when still practically unaltered by the incursion of the Europeans. True, I saw the occasional young man in the villages wearing the tin mining helmets then in use in the copper mines, and with a pair of boots on his feet even if he had very little else. To acquire these, he would have been to the Copper-belt, hundreds of miles away, and spent a year earning a trifle of the white man's money. I saw some women wearing dresses made of the brilliantly coloured Lancashire or Indian cloths that were imported and a few of the men wore khaki shirts and shorts obtained from the same source.

The Upper Zambezi floods over a wide floodplain every year after the rains. Of course, this annual flooding brought great fertility to the flood-plain and when the cattle returned after the water subsided they had plenty of grass to eat. To the Barotse people, as important as the cattle were the fields. The men looked after the cattle and the women hoed, planted and harvested the fields. In the higher country, beyond the floodplain, the people lived during the floods in temporary villages and grazed the cattle until the waters receded and they could return to the lush grass below. And there, also, were other people who practised slash-and-burn agriculture and grew corn and kept cattle, too. I travelled all over the country in the course of my work – inoculating cattle – and I never saw any hungry people, never came to a village without plenty of grain stored in it to make the stiff porridge and beer the people lived on, and I never even heard any stories about past hunger or famine. I never saw any signs of soil degradation or damage and I am quite sure that that particular form of agriculture could have continued into the future as long as the Zambezi flowed. All living things in Barotseland (and in most other parts of the savannah country of Africa too), formed a balanced soil community which, with its checks and balances, could have gone on indefinitely.

The small population partly accounted for this. It may have been kept in check by the lack of medical facilities but the impression I had was that it was due to a form of birth control practised by all the people. Mothers suckled their babies for up to four years and during that time they were, by unbreakable taboo, absolutely out of bounds to their husbands! I had only to consider the families that I knew well to realise the efficacy of this arrangement – the children of nearly every family I considered were well spaced out.

The other reason for the stable agriculture was that with iron-age tools and implements the people were quite incapable of destroying other forms of life too savagely, or altering the balance of nature too drastically. It takes a long time to cut down a tree with an iron axe. With a chain-saw, this takes only a few seconds. To kill wild animals with a spear or bow and arrow is generally so difficult and time-consuming that it is hardly worthwhile. With a rifle it is only too easy. You cannot destroy soil with slash-and-burn agriculture – unless you have access to chemical fertilisers – because long before the soil destruction point is reached you cease to obtain crops that will give you an increase on your seed input. Given chemicals, you can go on exploiting the land until it has nearly all washed away.

Now, with the advent of Europeans introducing their customs, old tribal traditions have begun to break down. The old taboos which prevented having too many babies have collapsed and so overpopulation has started. An inhabitant of Lealui I met recently told me that the population is now exploding: the great forests, which I thought of as eternal, are being speedily destroyed, the sandy soil of the higher country is being eroded and the wild game, of course, is almost gone. The soil community of that part of Africa has been thrown out of balance and will ultimately be destroyed.

The more tractors, chain-saws and agricultural chemicals that are imported into that country, the quicker will be the death of the soil with all the consequences that implies for the population.

The true effect of the death of the soil that is happening in so many parts of Africa, and in every other continent too, is currently masked by food aid – food sent as charity by other countries where there is a surplus. Western Europe and North America are the present providers of this surplus. If anything happens to *their* soil communities, the result will be global catastrophe on an appalling scale.

It is a mistake to suppose, however, that soil death is a new phenomenon on this planet so to find out about earlier examples of it I went, with the BBC crew for the television series, to the village of Magzalial, in northern Iraq.

In October 1983, we drove north from Baghdad to the city of Mosul. Mosul is hard by the ruined walls of Nineveh which was once one of the great cities of the world. But we were out to see the remnants of human life that flourished far earlier than Nineveh and we drove out into the desert, north-westward, to find the remains of what the archaeologists excavating it believe to contain the earliest traces of agriculture yet discovered on this planet.

We left, in due course, the abominable road along which we had been travelling and started along a dirt track. The land was flat and rolling, and in places we saw 'tells', which are the mounds created by the towns or villages of the past, each settlement having been built on the ruins of its predecessor so that the mud bricks used in each rebuilding broke down to form the base of a growing hill. It is in these tells that archaeologists love to excavate, for by doing so they dig through layer after layer of the remains of human settlements – and who knows what they may find?

There was a strong, hot wind and, in a growing dust storm, we saw modern farmers preparing to sow wheat and barley seed in anticipation of the coming rains. The country we were in was much higher than the alluvial plain of Mesopotamia and had a fair, though unreliable, rainfall which, perhaps three years out of four, would make possible very scanty crops of wheat. The farmers had tractors and combine harvesters, paid for by the money which had been earnt by Iraq's oil. It was quite obvious from what I saw of the landscape, and from what my Iraqi guide told me, that this dry-land agriculture would not be sustainable in the long term: the soil, laid bare by the tractor-drawn ploughs, was blowing away before our eyes! As the wind increased and the dust storm intensified, we could actually see the topsoil being lifted and blown away.

Besides the ploughed land there was much semi-desert with sparse shrub growing on it. We saw several large flocks of sheep and goats guarded by young boys, some of whom carried sub-machine-guns. It was quite obvious that the scrub that provided the livelihood for the livestock was heavily overgrazed. I was reminded of the grazing lands that I had seen destroyed by the flocks of karakul sheep in Namibia. The superabundance of grazing mouths damages the scrub and can eventually destroy it altogether and turn it into a desert.

The dust storm grew to such proportions that our drivers could hardly see their way at all. I reflected that pollen analysis had shown that all this country once was a hardwood forest. It is now inconceivable that any tree could grow in it at all.

We eventually found a village of mud-brick huts and a guide emerged from one of the latter, his face muffled by a cloth against the cutting grit and dust. Guide or not, we felt lost, the dust storm having attained a ferocity that made driving practically impossible, but we *did* eventually reach Magzalial: a small modern village, miserable and desolate, at the foot of a steep-sided natural hill. Below the hill on the other side was the deeply eroded gorge of a river where, surprisingly, open water was to be seen.

The modern inhabitants of Magzalial, about two dozen in number, warned us not to go to the very top of the hill because it was their graveyard. But we found the excavated remains of the ancient village: perhaps one of the most ancient villages in the world.

As I stood there, in the lee of an earth bank and braced against the wind, my eyes full of grit, I looked over the howling wilderness and remembered what I had read and been told of the history of the place.

The earliest villages in the high country of northern Iraq are now believed to have been built before agriculture was invented. The country was fecund enough to support human settlements of a permanent nature. I thought of those modern hunter-gatherers, the bushmen. They can never settle because their semi-desert environment is not supportive enough for them to be able to stay long in one place. They must constantly travel in search of food. But here, in these uplands, the forest was sufficiently rich in edible things for people to be able to settle in villages even before they had learned to dig or to plough. Here, if anywhere on Earth, was the garden of Eden.

But in Magzalial evidence has been unearthed that true agriculture was practised there and a very long time ago. About 7000 BC, there were settled farmers in the north of Mesopotamia. They were settled and farming there long before the most ancient cities were built in Mesopotamia or Egypt or indeed in any other part of the planet.

Besides grain and pulses, tree nuts and fruit have been found. Plentiful bones of many kinds of wild animals have been discovered in such villages.

Again, in this country few wild animals can exist. I tried to imagine the countryside as it must have looked 9000 years ago – shortly after the retreat of the last Ice Age. It would have been wooded, with oak, pistachio, almond trees and probably many others; the ground would have been green beneath the trees because the mini-climate of a wooded area favours grass and herbaceous plants. It would have been comparatively cool and humid because the not-inconsiderable rainfall of that country, instead of all running away to the sea in gulleys and rivers, would have sunk into the well-covered ground and been taken up by the plants and trees and transpired back into the atmosphere. The river below us, instead of being virtually dry, would have been steadily flowing with clear water. There is nothing like woodland for moderating and regulating the movement of water and, without soil erosion, the water would have remained clear. Café-au-lait coloured rivers are a sure sign of soil erosion. The people of the village would be engaged in collecting nuts and other tree fruit, hunting wild animals, fishing in the river. Some of the women would have been hoeing small clearings of land to sow the seed of newly domesticated wheat and barley. Much of their stored grain would have been from wild grasses. Maybe some of the men and little boys would have been caring for small flocks and herds of newly-domesticated animals. Those people could never have imagined the howling wilderness, with its drought and dust storms, in which their successors drag out a miserable existence today.

Progress can mean several things. R.J.Braidwood, the archaeologist who excavated the very similar Neolithic village of Jarmo during the 1950s, wrote:

During the period which has elapsed since Jarmo was a village, *man* has been the pre-eminent environmental influence, and the effects of his handiwork are to be seen throughout the Near East. In general, the role of man, of his agriculture and of his flocks has been destructive, and this without any one man wishing to be destructive.

... throughout much of the once-wooded plain and foothill area of the Chemchemal valley, hardly a shrub remains. The scrub-oak is rarely allowed to reach more than six feet in height before it is hacked away by the charcoal-burners. With the trees and the bush cover gone, and the grass eaten down to its roots each spring, the soil has largely gone too to silt up the rivers ... In winter it washes away on every slope almost as fast as it can form and the rains rage off the land in chocolate torrents.

After seeing the evidence of the earliest village settlements, we decided to look at the earliest cities. We drove south along the long, long road from Baghdad to Kut, where a British army suffered and mostly died during the First World War, and there turned right and drove to Nasiriya, at which small town we arrived in the evening. On the road we had passed a massive convoy of heavy tanks, loaded on transporters, heading north, to counter an Iranian offensive near Sulaymaniya, for Iran and Iraq were at war.

We refreshed ourselves after the gruelling drive by strolling along the bank of the Euphrates in the comparative cool of the evening. Close to Nasiriya, south-west, is Ur of the Chaldees, but we could not go there because the site had been turned into a rocket base. So, next day, we drove north-west instead.

We drove for some time along marshes and lagoons that border the Euphrates, and had just this short glimpse of the strange amphibious world of the Marsh Arabs. Then the track, for it was no more, took us out into the desert. Much of this was snowy white and glaring in the very hot sun. The whiteness was due to salt. Vast areas of that country are salted thus and much of it has been since the Sumerian civilisation collapsed, indeed it is speculated that salinisation of the soil was the main cause of that civilisation's decline.

I have seen salinisation on a vast scale in the Valley of the Indus, in Pakistan, and I have seen its beginnings in California. It seems inevitable that if irrigation is abused in a dry, hot climate, salinisation will result. The water soaks down into the earth, dissolves any soluble chemical which may be there, such as salt, climbs up to the surface again by capillary action and is evaporated leaving its mineral load behind it, on the surface. Very good drainage can contain it, and even sweeten salted land. Water is allowed to flood the land, it sinks down and is carried away to the river again by underground drains. I have read that the Sumerians knew all about this, knew how to prevent it and even how to cure it but that the greed and short-sightedness of their farmers precluded them from taking the remedies. After all – then as now – if a farmer can make a big profit *for the next ten years* by taking a certain course he will do that no matter what effect his actions have on the generations that will come after him. It is too much to expect a poor farmer to expend great sums of energy or money to desalinise his land for the benefit of posterity when he can make a profit by growing, with heavy irrigation, a crop *now*. The Sumerian farmer was no doubt just as deeply in debt as our own farmers are today. Posterity was not going to pay the interest. Then, over more howling desert and salinised soil, we came to Uruk, Warkha in Arabic.

The sun was hot by then and the glare from the desert dust was blinding. We saw, ahead of us, some hummocky desert and came to a stop in between some low, flat-roofed, mud-brick buildings. Nearby a small gelding stood tethered to the one tiny tree for many miles. He had a beautiful Arab saddle on his back.

Numerous small children peered shyly out of the big, low building on one side. Out came a man, dressed traditionally in a *cheffiyeh*: the Arab robe. His thin, smiling face was blackened by the sun. He was the caretaker, employed by the team of German archaeologists which was currently excavating the site. The latter were not there as digging takes place only in the winter, when it is cooler. Except for the man, and his wives and children, Warkha was inhabited only by ghosts.

We trekked across the hot, dusty ground. This was littered with pot-sherds – pieces of broken pottery. Everything was the same colour – pottery, bricks, ground – a dusty light-brown. The extent of the disturbed and raised ground was enormous and anywhere in it you could hardly avoid stepping on a piece of broken pottery. I reflected that below our feet was layer after layer of city foundations, burnt bricks and pottery – city under city under city. For 2000 years, as long as from the birth of Christ until now, a city, large even by modern standards, existed at Warkha. From its founding, probably in about 4000 BC, to its decline in about 2000 BC, generation after generation of townspeople lived and died, farmed the land, fished in the rivers, hunted the marshes and forests. If this was not the first city to exist on planet Earth it was certainly one of the first, and the earliest writing ever discovered may have been found at Warkha.

As we walked, feeling hotter and hotter and thirstier and thirstier, we came across the uncovered remains of many houses and temples. The latter had mud walls decorated with mosaics composed of thousands of small clay cones with decorated heads. We continued on towards the ziggurat: the great temple that formed the centre of the city. Temple on temple was built: as one temple became old or old-fashioned it was razed and a newer one built upon its rubble. The temples were built high, in an attempt to bridge the gap between this world and that of the Gods.

From the top we could look over the entire extent of that part of the city which is contained within a wall. But outside the wall there were suburbs, or more city, and I could see another ziggurat several miles away. The Iraqi archaeologist who was attached to us said it had been part of the same city. I reflected that it would have taken an hour to have walked to it. Nobody knows how many people lived in that city at any one time but it is known that many, many people lived there for at least 2000 years. The founda-tions of the houses show that it was a crowded city. I reflected that the high temple on which we were standing, and most of the houses, too, were built of burnt brick, and burnt also was the pottery that had provided that carpet, square miles in extent, of shards. Where did the trees come from that provided the fuel for all that firing? From the lands around the headwaters of the Euphrates perhaps? Floated down the river? Maybe this great city, and the others like it – Ur, Eridu, Al Ubaid, Babylon and the rest of them – proved too great a weight for the forests of Armenia and Azerbaijan to bear? Maybe that is why those countries, which have such rich pollen counts indicating that they had once been forested, are such desolate and unproductive regions now?

I wished I could have had, standing next to me, a Sumerian farmer from about 2500 BC. I wished I could have seen his expression as he looked upon that blasted ruin and the miles of desert about it. I would have liked to have asked him, 'Did you expect *this* to happen?' And if a living Cambridgeshire farmer, or a Kansas farmer, could travel into the future and stand in the middle of his farm in, say, 1000 years, what would he see? Could it be that, those millennia ago, the Sumerian farmers were as confident in the rightness of their methods as the farmers of Cambridgeshire and Kansas are today? Things happen faster now. There is nothing like standing on the ruins of the distant past to make one wonder what the future will hold.

Above, *Existing hunter-gatherers such as the Bushmen of the Namib and Kalahari Deserts feed themselves well from their environment;* below, *Pygmies from Zaire smoking tobacco*

41

Opposite, *Fishermen of Lake Turkana in northern Kenya do not overdeplete the stocks of fish;* above, *Empty ostrich eggs are used to store water;* right, *Bushmen manage to find enough food for their families even in the inhospitable Kalahari Desert*

Above, *The Masai are nomadic cattle-herders;* left, *A Pygmy from the Congo forests makes fire by friction;* opposite above, *Decaying machinery in the failed irrigation project at Malka Dakaa in Kenya;* opposite below, *Tea is a cash crop in Kenya, sold for foreign currency needed to buy luxuries*

What Has Happened to the Grecian Soil?
———————— John Seymour ————————

Some time between 1700 and 1400 BC a significant change was made in the burial customs of the people of Crete. Hitherto, the dead had been buried in wooden coffins. After the change they were buried in earthenware ones. Potters will say that any inference drawn from this that the change was due to an increasing shortage of trees on the island is invalid because pottery takes wood to fire it. The answer to this is that you can fire pottery, as Cretan potters do today, with very small wood. You can only make proper coffins with large planks, which are cut out of large trees. Also, the pillars which held up many palaces of the Minoans were made from huge cedars. You would be hard put to find a cedar, or any other tree, big enough to make one such pillar on the island now.

Pollen analysis and all the archaeological evidence indicate that the lands around the Mediterranean basin were once well-wooded and fertile. You cannot say that about them now.

It was late autumn when we filmed at Knossos and the air was cool although the sun was brilliant. The ruins of the palace – if palace it was, it seemed more like a town to me – are impressive and beautiful but you have to go to the magnificent museum in Heraklion to see the true glory and splendour of the Minoan civilisation.

People were cultivating the ground for wheat in Crete with stone tools 8000 years ago from now. At that time, they learnt to make pottery and bake bricks. Bronze came in, with a new incursion of people, about 4600 years ago. These people were the founders of a civilisation that some might say (and I would be among them) has never been surpassed.

The Minoans were a people without armies or land-based weapons. They relied for their defence on their unrivalled command of the sea. Their cities needed no walls to defend them. They were a nature-worshipping people who worshipped the Goddess rather than the God, and every artefact they made which survives them points to a close and loving relationship with the rest of nature on their beautiful island and the sea around it.

Evidently they began to be challenged on the sea by the middle of the

Previous page, *A Chagga homegarden on the slopes of Mount Kilimanjaro. The Chagga people's agro-forestry system is considered a model form of land use for tropical environments;* opposite, *Cariba Dam in Zimbabwe. Giant dam projects are no longer considered to be the 'grand solution' to the Third World's problem of underdevelopment as the social and environmental costs have often proved prohibitive* fifteenth century BC. For the first time we find that weapons were placed in the tombs of the illustrious dead found among ruins dating from this period. In the middle of that century a terrible disaster came upon them. Some people attribute this to the volcanic explosion on the Island of Thera, with its tidal waves and earthquakes, but there had been devastating earthquakes before on the island and the towns and palaces had been rebuilt. Certainly, whatever the disaster was, the Minoans never recovered from it. Their civilisation quickly decayed and the island was overwhelmed by the far less civilised Achaeans who came over from the Greek mainland.

Karl Marx interpreted all human history as being the product of economic forces only. The Reading Room of the British Museum Library is very far removed from the soil. The human population of Crete was very large in Minoan times: Arthur Evans, who excavated Knossos in the two years following 1900, estimated that there were 80,000 people in Knossos alone and there were many other palaces, villas and towns all over the island then. Crete was an exporter of food at that time: the gold, copper, precious stones and other valuable raw materials found so copiously in the ruins were all bought by the export of food. I contend that the Minoans lost their *soil*, and that it was that which weakened and finally destroyed their civilisation: the volcanic explosion was just the final *coup de grâce*.

Vistagi is a mountain village, like scores of others, quite high up on the southern slopes of the mountains that form the background to Crete. The sheep of the village graze the high slopes of Mount Ida, the mountain with the cave in which Zeus was born.

The drive south from Heraklion across the island gives an idea of what has happened since Minoan times. There are pockets of soil in low-lying places, for the soil which has eroded from the uplands has to go somewhere. Most of it ends in the sea but some of it is deposited in valleys and low places and now forms the basis of profitable gardening and farming. These isolated patches give an idea of what must have been the extreme fertility of the island in days gone by when soil, held by trees, must have covered all but the highest peaks.

We leave the good coast road to cut inland up to a stone-paved road provided with many and grievous potholes and after a long and rough drive we reach Vistagi.

Although the soil has been eroded, the hillsides are not completely treeless. They are lightly clad with the sort of trees that grow in poor soil: olives, carobs, almonds, figs and, in the better places, walnuts and citrus fruits. Between these trees are aromatic herbs in great profusion: tough, woody, drought-resistant shrubs which are hardy enough to withstand constant grazing. For the sheep and the goats are the scourge of the soil in all Mediterranean lands: they make quite sure that there is no chance of the soil returning.

Soil is created from rock. The rate of soil creation differs all over the world, according to what rock it is, in what climate and a hundred other factors; but North American estimates put soil creation at between one inch of topsoil per 300 years and one inch in 1000 years. Certainly it is a slow process.

Not so soil erosion. I have seen an erosion gulley in Kansas that you could hide a two-storey house in which was created in one year. But the roots of the olive trees, which make up the preponderant flora here, bind what soil there is, find their way down into fissures and cracks of the rock and no doubt help to grind away at and shatter the rock to help it on its way to become soil. Without the olive these mountain slopes would very quickly revert to bare rock.

Down in a valley we came upon what to me was a hideous sight. Bulldozers were busy tearing out the olive trees – tearing up their roots and exposing the shallow soil beneath them – because some rich man wished to farm this land for a few years with cash crops and make a lot of money. But, as anybody could see, and as the man must have known very well himself, he was destroying for ever that part of his country. He may have cash-cropped it for ten or a dozen years. He may have become rich. But he would have left barren rock to his descendants.

As we approached Vistagi we found small terraces. These were true terraces, that is they were level, or nearly level, areas of soil held up on the hillside by well-built rock walls. As long as men maintain the walls, and properly farm the terraces, the soil will remain there, produce good crops and even be improved. If the rock walls were neglected for even a few years, they would fall down and the soil would disappear within a short time.

But the terraces were well maintained and the soil in them was good. Yet the terraces were tiny. Most of them were the size of a large room. A farmer in Kansas or East Anglia would be amazed that anyone could think it worthwhile to plant wheat in such tiny areas of land.

We approached Vistagi. Now whatever I can write about the *hospitality* that we received in this village is bound to be an understatement. Nothing can describe it. It started modestly, with the Mayor opening the proceedings with tiny cups of sweet coffee accompanied by glasses of *raki*, the local firewater. We spent a week there and soon realised that practically every day was a feast day of some sort or another. Our first feast day consisted of our winding our way up to the top of the village, up a path so steep that it had to have steps cut into it, to the house of an old soldier who lived there.

Inside the house was a large table – absolutely laden with food. The food, nearly all home-grown, consisted of huge platters or bowls of mutton grilled on charcoal, snails, olives, chicken, walnuts dipped in honey, almonds and other knick-knacks that I have probably forgotten. Drink consisted of wine, more wine and more wine, and *raki*.

What we did not realise on the initial feast day was that our first port of call, to the house of the old soldier was not to be, by any means, the last. After eating and drinking to absolute repletion we all fared out of that house and began the descent of the steep street. And lo and behold, in we went to *another* house and there *another* table was laid out, if anything more lavishly, and again the process was repeated. And there was *another* house and *another* house! My counting faculties were perhaps impaired somewhat by all this hospitality but I do seem to remember *seven* such houses on one hard-fought day.

We went to Vistagi to observe a village of peasants who were still living in the age-old peasant tradition of almost complete self-sufficiency. I describe this lavish entertainment to indicate that people living thus are the *really* wealthy people of this globe. These people had more than enough of every single thing they really wanted: the only foodstuff they had to buy was coffee and, if they really felt they needed it, fish. The coffee was imported and cost hard cash: the fish was caught by fellow-peasants of the sea and these would take the country produce in exchange.

There was an olive oil mill in the middle of the village. Every house-holder made his own wine. Every man had a terrace or two in which he grew wheat. The land was ploughed by donkey-pulled ploughs and har-rowed likewise. The seed was broadcast by hand. The simple steel com-ponents of the ploughs were made by a blacksmith in a village very near. The sheath knives, or daggers, that every man wore stuck in his belt were made by an old man in another village over the mountain. Every woman, on marriage, had a loom given to her and on it she would weave all the cloth for the family.

And how do such obviously well-fed communities exist on land from which most of the soil has gone? Well the vine, the olive tree, and arable crops grown *strictly* within the confines of terrace walls provide the answer. If the Minoans, 4000 years ago, had had the art of terracing and practised it, and if they had had more respect for trees and spared their hillside forests, that country would now be able to support a much larger population, and export wheat as the Minoans did.

It has been said, and probably rightly, that the Athenians owed their pre-eminence in the arts, sciences and commerce to the poverty of their soil. When their soil went, under the assault of too 'efficient' bronze-age farmers, they were forced to import wheat, which they could only do by exploiting their *subsoil* with the olive and the grape: those crops of the subsoil and eroded land. They paid for the wheat with wine and oil. To transport this they had to build ships and make amphorae. So their industries developed. Their seafaring abilities improved. They were forced to trade all over the Mediterranean world – and colonise as well. But all such societies – societies that depend on the soil of other people – eventually come to an end. Let us hope that the citizens of Vistagi, and all the other villages like it, look after the little soil they have left. For they are creatures of the soil as we all are and, if that goes, they go too.

Concern about loss of soil in Greece was expressed by Plato as long as 2400 years ago in his book *Critias*. His descriptions of the impact of deforestation and farming on *Attica* are as vivid today as those of any contemporary campaigning ecologist:

All other lands were surpassed by ours in goodness of soil, so that it was actually able at that period to support a large host which was exempt from the labours of husbandry. And of its goodness a strong proof is this: what is now left of our soil rivals any other in being all-productive and abundant in crops and rich in pasturage for all kinds of cattle; and at that period, in addition to their fine quality, it produced these things in vast quantity . . . And, just as happens in small islands, what now remains compared with what then existed is like the skeleton of a sick man, all the fat and soft earth having wasted away, and only the bare framework of the land being left. But at that epoch the country was unimpaired, and for its mountains it had high arable hills, and in place of the swamps as they are now called, it contained plains full of rich soil; and it had much forest-land in its mountains, of which there are visible signs even to this day; for there are some mountains which now have nothing but food for bees, but they had trees no very long time ago, and the rafters from those felled there to roof the largest buildings are still sound. And besides, there were many lofty trees of cultivated species; and it produced boundless pasturage for flocks. Moreover, it was enriched by the yearly rains from Zeus, which were not lost to it, as now, by flowing from the bare land into the sea; but the soil it had was deep, and therein it received the water, storing it up in the retentive loamy soil; and by drawing off into the hollows from the heights the water that was there absorbed, it provided all the various districts with abundant supplies of springwater and streams, whereof the shrines which still remain even now, at the spots where the fountains formerly existed, are signs which testify that our present description of the land is true.

Such, then, was the natural condition of the rest of the country, and it was ornamented as you would expect from genuine husbandmen who made husbandry their sole task, and who were also men of great taste and of native talent, and possessed of most excellent land and a great abundance of water, and also, above the land, a climate of most happily tempered seasons.

Rome and the Soil
— Herbert Girardet —

The Mediterranean basin was the cradle of modern urban civilisation. By the time Rome was founded, around 750 BC, dozens of cities had sprung up on the shores of this 'inland ocean' which links Europe, Asia and Africa. Trading ships were sailing back and forth between Athens, Tyre, Carthage, Troy and Memphis, exchanging African gold and ivory for Greek pottery and olive oil, Lebanese timber for Asian bronze implements. Warships were often on the move from one city, seeking to challenge the maritime power of another. In war and in peace, people and ideas found their way from one continent and one urban centre to the next.

As populations grew, as cities were destroyed, as lands were denuded and spoiled, tribes swarmed and went to other places. The Cretans settled all around the coasts of the Mediterranean. The Phoenicians ventured out from their base at Tyre and started a powerful new city in Carthage. The Greeks took over parts of Italy and also set up colonies in Asia Minor and North Africa. The Trojans, after their city was burnt, criss-crossed the Mediterranean until they succeeded in putting down roots again in Latium, on the banks of the River Tiber. According to Virgil, Aeneas and his fellow Trojans became involved in fierce battles until they finally managed to make an uneasy peace with the inhabitants of Latium with whom they were ultimately to merge. Together they founded the ancient world's most powerful city: Rome.

Italy, like most of the lands on the shores of the Mediterranean, was blessed with a well-tempered climate and was a place of great fertility, much of it covered with volcanic soil and clad with deciduous forest. There was an abundance of wildlife, several perennial rivers and plenty of fish in inland waters and the sea. Wherever villages sprang up forests were converted either into open woodlands to provide pasture for cattle, sheep and goats, or into fields for growing grain and other crops. As settlements grew, more land had to be taken over to feed their populations or food production had to be intensified on the available space. Some towns started to specialise in the manufacture of useful articles like metal implements, pottery or woollen garments which enabled them to buy in foodstuffs from other places in exchange.

The Roman Republic was founded around 500 BC. By this time the Romans had utilised nearly all the potentially cultivable land that was part of their territory. They had carried out extensive drainage works, the remains of which can still be seen today. The early Roman Republic controlled only a small area of some hundreds of square miles which held a population of several hundred thousand people. A population density of between 500 and 1000 people per square mile has been suggested for this period. This would have made intensive agriculture necessary. Latium – the region surrounding Rome – was itself quite densely populated by other tribes, and since Rome did not have a significant manufacturing base at this time she would have been unable to buy in sufficient quantities of food from elsewhere. Barley, spelt and millet were the main staple crops of Rome during the early days of the Republic. The people were probably mainly vegetarian though fish from both the river and the sea must have made quite an important contribution to their diet.

The first hundred years of the Republic were spent consolidating and then expanding Rome's territory at the expense of neighbouring tribes. By 270 BC virtually all of Italy except the Po valley had been taken over by Rome. Her astounding success was partly due to the policy of not killing or enslaving all the inhabitants of the conquered lands, as was the usual practice at that time.

Rome's rapid expansion in Italy became a cause of great concern for the other major power in the Mediterranean of the time, Carthage. The Phoenicians who had founded a powerful colony in the North African coast at around 800 BC, had control over Sicily and Sardinia. In the First Punic War (265–241 BC), the Romans succeeded in taking over Sicily. Sardinia and Corsica were merged with the Roman empire soon after. Then Rome continued to enlarge her territory by annexing the fertile Po valley. Finally, in the Third Punic War (153–146 BC), Rome conquered Carthage and with it her extensive North African territories. This was not the end of Rome's expansion. Roman armies subsequently subdued Macedonia in a brief war. Greece, too, came into Rome's fold. Then Syria was subdued, followed by much of the rest of Asia Minor. Egypt was soon incorporated in the Roman empire, and so were Spain, Gaul and Britain. All this took place in the course of about 400 years. The most powerful empire ever assembled until that time was put together under the auspices of the *'pax Romana'*. It was a staggering achievement of willpower and determination to dominate.

The Romans did not only set out to dominate the populations of territories which they absorbed within their empire, they were above all else concerned with gaining access to the resources of these regions. And the most important resource was the soil. In the early days of the Roman Republic the *ager Romanus*, the soil of Rome, just managed to feed its population. Most people were involved in food production, at least during harvest time. Peasant farming, that is, small-scale farming for local need, was the predominant mode of food procurement. During Rome's expansion, increasing numbers of loyal citizens were needed as soldiers and thus ever larger numbers of peasants had to join the army. The consolidation of Rome's power over larger and larger territories enabled her to extract tribute in the form of foodstuffs – mostly grain – from these conquered regions. At the same time, slaves were brought back from the colonies to work the *ager Romanus*. Much of the readily accessible land became the property of rich Roman citizens. One such landowner was Cato, who grew up on a farm and for whom farming was a passion all his life, even after having distinguished himself as a soldier, historian, orator and legal expert. His extensive landholdings were mostly worked by slaves.

Cato's book *De Agri Cultura* gives detailed instructions to the would-be gentleman farmer on how to set up a large mixed farm on which to grow grapes, olives, timber, fruit, grass and cereals. It discusses issues such as soil, climate and location. It is concerned with windbreaks and with maintaining soil fertility by the use of the dung of goats, cattle, sheep and pigeons, and other organic wastes. It gives instructions on how to make compost from straw, lupins, chaff, beanstalks, leaves and other plant materials. It also gives veterinary advice and even has a section on naturopathic medicines for humans.

Reading *De Agri Cultura* in the original, one cannot help being struck by its similarity, in some respects, to a modern manual of purely cost-conscious industrial farming. Cato . . . was a soldier primarily, and a very large landowner, and his attitude is especially one of an employer instructing employees in order to get the biggest profits for himself out of the operation . . . He even notes that troublesome slaves should be worked chained together . . . Yet everything Cato recommends is aimed at maintaining the natural biocycle, at putting back into the soil the residues of that which had been taken from it in order to maintain its fertility. How far artificial chemical fertilisers do this today is a nice question. (A.H.Walters, *Ecology, Food and Civilisation.*)

It is unlikely that all the Roman noblemen who became the great land-owners of Italy were as concerned as Cato with careful husbandry and maintenance of soil fertility. It was a status symbol to be the owner or leaseholder of an estate. The richest men were usually also the largest landowners and proprietors of slaves. Roman agricultural slaves were used in much the same way as the slaves on the estates and plantations of the modern world: to provide muscle-power to keep agricultural production going. Huge numbers of slaves were brought back by Roman armies from their expeditions around the Mediterranean:

As a result of the victorious campaigns of the third and second centuries BC, slaves flooded into Rome: 75,000 as prisoners of the First Punic War . . . 30,000 from Tarentum alone, among the numerous captives of the Hannibalic War; huge numbers of Asiatics after the victory over Antiochus III; 150,000 from Epirus in 167 BC. They were sold in the great slavemarkets of Capua and Delos. These enclosures were capable of handling 20,000 slaves a day. (M.Grant, *History of Rome.*)

It was the slaves who made the great estates and plantations prosper and who made their rich owners a great deal richer. These estates, the so-called latifundia, dominated the lowlands of Italy after the Punic Wars. They produced the meat, fruit, wine, olive oil and vegetables consumed in Rome and other major cities. Grain, as we shall see, was increasingly produced elsewhere. In the hills, small-scale subsistence farming did continue, though no doubt under considerable pressure. It is clear that the never-ending wars deprived farming families of male labour and in Roman law land which lay neglected could legally be claimed by anyone. Only the rich could afford the slaves to work the land. It was only the more remote land in the hills which was considered less desirable by potential urban land-owners. So here there was some continuity of small-scale landownership even if the land had to be left neglected for periods of time.

While the main change in the lowlands was the disappearance of the small farms in favour of large estates worked by slaves, the main impact of Roman expansion in the hills and mountains was on vegetation cover. Deforestation of these regions took place on a massive scale. Timber was required for house construction in Rome and other building works, but even more so for shipbuilding. The growth of Rome's Mediterranean empire required a large fleet of ships for both trade and war. The Apennine mountains were denuded of trees during the Punic Wars. Whether con-scious attempts at replanting were made later on we do not know, but even in times of peace – which were rare indeed – demand for firewood and timber must have continued.

Large quantities of soil must have been washed down from the hills, silting up rivers and lakes and contributing to the spread of marshes in the coastal areas. The Pontine Marshes near the mouth of the River Tiber were formed towards the end of the Punic Wars and this previously very productive region could not be used for farming to any significant extent until it was drained in the first decades of the twentieth century. The harbour at Paestum near Naples was completely silted up because of hillside deforestation and became unusable. Indeed, long stretches of the Italian coastline suffered from an accumulation of silt and from marshy conditions which date back to pre-Christian times.

Deforestation, loss of topsoil, the spread of swamps, the disappearance of the self-contained family farm and its replacement by latifundia worked by slaves . . . all of this had an impact on the capacity of the *ager Romanus* to feed the growing population of Italy. But as the Roman empire expanded, land for producing the staple crop, wheat, became available elsewhere. Sicily became a major source of wheat for Rome after it was annexed during the First Punic War. At around 200 BC, when Carthage was taken over by Rome, the soil of north Africa became accessible to the

Romans for the first time. To start with it was only the area around Carthage itself, but in the course of a few decades much of the land of north Africa above the Sahara was under Roman control. By 30 BC Rome's power reached right across to Mauritania where Roman colonies were founded. Rome was to rule north Africa for more than 500 years. Nearly 600 towns and cities were built there and today nearly all of them are ghost towns.

North Africa, too, had substantial forest cover in pre-Roman days. Carthage itself was primarily a trading city in the old-established Phoenician tradition but it had also developed an effective farming system for supplying staple crops. The Romans had great respect for Carthaginian land management. The Roman senate had the agricultural writings of Mago, famous for his detailed analysis of cultivation techniques and land use in Carthage, translated into Latin.

Much of north Africa was controlled by pastoralist tribes when the Romans first arrived there. Some of them seem to have been loosely organised in a nomadic pattern of life, others were more tightly structured under the central authority of chieftains and kings. Ancient travel-writers stress that farming was also practised in some areas by native tribes. There is little doubt that the relationship between nomadic pastoralist groups and farming communities was similar to that still found in east Africa until quite recently. There tribes like the Tutsi, who were cattle-, sheep- and goat-herding pastoralists, had settled farming tribes, like the Hutu, 'in their care' who would supply them with grain and whom they were supposed to protect in times of warfare. This sort of arrangement was common all over east Africa until the colonial powers in the nineteenth century teamed up with farming tribes and thus weakened the almost feudal power the pastoralists had over them.

For Rome, the newly-conquered African territory became a vital source of grain which was to supply the capital city for several centuries. The Romans introduced their methods of intensive cultivation to increase the crop yields. Many veterans were settled in north Africa to farm the land and to intermarry with local women. The newly-built ports such as Leptis Magna were used for grain shipments and for trade in gold and ivory from the African interior. Leptis was only three days' sailing away from Rome.

It was Julius Caesar who realised the full potential of north Africa's soil and he extended Rome's rule further west to include much of the land area north of the Sahara. He continued the policy of settling Roman war veterans there. Some of them were given their own land, others worked as sharecroppers on land owned by rich Romans. Nearly 50,000 tons of grain were produced each year. One hundred years later, Africa supplied two-thirds of the wheat consumed by Rome, 500,000 tons annually.

At this time most of the suitable land in north-west Africa was used for arable farming. Pastures and orchards gave way to the plough. The cheap food policy which had been adopted in Rome to keep the poor of the city contented required the import of grain at the cheapest possible price. The pressure of Roman farming led to the expansion of arable land into the hill-country. Forests were removed to make room for farms. Extensive terracing was undertaken to prevent erosion. In suitable areas olive groves were planted, for Rome's demand for olive oil was increasing all the time. These activities were at the expense of the traditional pastoralist life-style of the Berber tribes who were forced to adapt to a more settled existence.

The Romans built extensive irrigation systems in north Africa. In some areas they 'terraced hillsides not for cultivation but to hold rainfall or melting snow long enough for it to sink in gradually and fill the water-table for wells'. (S.Raven, *Rome in Africa*). Artificial basins and cisterns were built for the same purpose. There is no doubt that care was taken in some areas to preserve the fertility of the land, to preserve scarce water resources and to prevent soil erosion. But in the end it was not enough.

Over hundreds of years the vegetation cover of north Africa was reduced still further. On much of the farmland only a crude form of crop rotation was practised by letting land lie fallow in alternate years. The sustained export of wheat also resulted in a reduction of soil fertility: much of the fertility of the land of north Africa went through the stomachs of Romans and the sewers of the city into the Mediterranean.

The mechanism of expansion of Roman power was a model which was later followed by other empire builders. It is worth examining in some detail.

As we have seen, Rome's power lay above all else in agriculture. As it expanded in Italy, more and more land was converted from forest into pasture and then into arable land. Each of these transformations resulted in higher levels of food production. The process of expansion required soldiers who were mostly recruited from the farms. Since Rome was nearly always engaged in a war somewhere, she needed a constant supply of soldiers. This permanent state of warfare undermined the existing farming system: farmers who are needed as soldiers necessarily have to neglect their land in their absence. The Roman law which required neglected land to be redistributed was a device which was used by the Roman nobility to their great advantage. It enabled them to accumulate large estates which they 'stocked' with slaves brought back from military expeditions.

Agriculture and warfare were the two most important economic activities. The first procured food, the second yielded loot. War booty provided the capital for setting up the latifundia, estates worked by slaves in Italy itself and increasingly in other food-producing areas. The peasant-soldiers of Italy had a stake in this system in so far as they received a share of war booty and also had the option of acquiring land of their own in newly-conquered areas. As we have seen, Roman veterans became the new farmers of north Africa. Not only did they produce food for Rome, they also were required to pay a land tax of between 25 and 33 per cent of their farm income to the Roman treasury.

From time to time attempts were made by Roman politicians to reverse the process by which more and more land in Italy fell into the hands of rich Roman absentee landlords but these were generally unsuccessful. The unwillingness of the owners and tenants of latifundia to give up their land holdings accelerated the process of making farmland in the colonies – notably in north Africa – available to Roman citizens. It was Julius Caesar who annexed large tracts of north Africa for Roman farmers. This policy made him very popular with the masses and assured a broadening of his power base. Hundreds of thousands of landless Roman peasants emigrated to the new territories to try their luck as sharecroppers.

Caesar had hoped that a large proportion of the urban poor would be siphoned off into the colonies. Keeping them supplied with food was a costly business of the state. Food for the urban dwellers had to be produced by muscle-power alone, both animal and human. No tractors or combine harvesters were available to work the land or bring in the harvest.

There is no doubt that the demand for food by the urban population was a great burden on the rural dwellers. In the outlying provinces in particular, like the territories of north Africa, there seems to have been a great deal of bitterness and resentment among the country folk against the landlords as well as urban racketeers who controlled the food trade:

Peasants certainly had reason to revolt. Consider the seizing of their stocks of food by profiteers in the very heart of famines and the wretched substitutes they had to fall back on; their deprivation even in good years, shown by their crowded housing, short life expectancy, exposure of children, and bitter disputes over the use of land and water; their indebtedness, while urban prosperity unfolded column-lined streets, public bath buildings and private palaces. (R.MacMullen, *Roman Social Relations 50 BC to AD 284.*)

Peasant revolts did break out repeatedly. Armies were sent to put them down and usually the country folk submitted after forceful displays of military power.

The Roman empire thrived on growth and expansion and as long as it could maintain this momentum it managed to sustain itself. With the reign of Emperor Augustus it almost reached the limits of its growth, though a few territories were acquired after his time, notably Britain.

After Augustus the slow but inexorable decline of Roman power got under way. Many reasons have been suggested for this, notably the rise of Christianity and the never-ending attempts by barbaric tribes like the

Goths and the Vandals to snatch territory from the clutches of the Romans. Whatever the reasons, the consequences for agricultural production were serious. It is apparent from a variety of historical sources that production did decline both in mainland Italy and in north Africa.

The philosopher and poet Lucretius observed around 60 BC that all was not well with the soil of Italy. He believed that the earth was dying, that the land was becoming exhausted, that the rain and the rivers were eroding it and carrying it down to the sea. He noticed that the farmers had to farm more land and to work harder than their ancestors to produce enough to support themselves.

In north Africa, around AD 250 St Cyprian, the bishop of Carthage, became pessimistic about the future. He wrote these words in a letter to the Roman proconsul of Africa:

You must know that the world has grown old and does not remain in its former vigour. It bears witness to its own decline. The rainfall and the sun's warmth are both diminishing; the metals are nearly exhausted; the husbandman is failing in his field . . . springs which once gushed forth liberally, now barely give a trickle of water. (V.G.Carter and T.Dale, *Topsoil and Civilisation.*)

For centuries Rome had extracted timber and grain from the land of north Africa. There is no doubt that loss of topsoil from deforestation and loss of humus and plant nutrients all contributed to the decline of soil fertility. After Rome's power in Africa collapsed as the Vandals invaded, agricultural land was once again turned into grazing land. Irrigation works collapsed and fell into disuse. Herds of cattle, sheep and goats prevented the regrowth of trees and forests. Lack of vegetation cover encouraged the spread of the desert further and further north.

It was the Roman writer and historian Pliny (AD 23–79) who expressed the conviction that the latifundia were the ruin of Italy. (*Latifundia perdidere Italiam.*) By this he presumably meant that they were the ruin of the Roman empire as a whole. There seems to be plenty of evidence to support this view.

As we have seen, the latifundia came about as a result of the conversion of the peasant into a soldier, and the annexation of neglected peasant land by would-be urban landowners. With the growth of the latifundia in Italy, north Africa, Sicily and elsewhere, agriculture became a purely profit-oriented enterprise. Fortunes were made from food production for the ever growing cities. This may have been of great benefit to the landowners but it had dire social and ecological consequences. Pliny stressed that he was appalled by the practice of using slaves straight from prison in the fields, since their work was very bad, as is everything done by men without hope. He felt that it was morally repugnant as well as economically disastrous.

With latifundia becoming the main agricultural unit throughout the

61

empire, the interest of the citizens in the land became purely commercial. The link between land and people was effectively broken and the long-term interest in its fertility was put aside. A number of peasant farms seem to have survived in less accessible regions, but the fertile soil of the lowlands of Italy was in the hands of absentee landowners. Slaves and disillusioned sharecroppers did little to keep it in good heart.

In addition to the disinterest of those who worked the land, the fertility that was taken out of it by transporting its produce to urban markets was never returned. City sewage – or, for that matter, artificial fertilisers or guano – did not find its way back to the fields. The very rudimentary crop rotation practised on Roman-owned fields did little to restore their fertility after harvests were taken out year after year.

The Romans were the first to test the viability of large-scale commercial agriculture. They could only make it work by gaining access to ever larger areas of land. But, clearly, in the end they recognised the limits to growth. They left much exhausted land behind them, as well as a trail of human tragedy. Here, in a passage from Virgil's *Eclogues*, the poet reformulates the experience of many a peasant farmer:

But the rest of us must go from here and be dispersed. To Scythia, to bone-dry Africa, the chalky spate of Oxus, even to Britain – that place cut off at the very world's end. And when shall I see my native land again? After long years? Or never? See the turf-dressed roof of my simple cottage and wondering gaze at the ears of corn that are all my kingdom. To think of some godless soldier owning my well-farmed fallow. A foreigner reaping these crops! To such a pass has civil dissension brought us. No more singing for me!

The Roman Inheritance
———— John Seymour ————

The mighty weight of Rome, as we have been told, destroyed a large part of the fragile soil of Italy, to say nothing of turning north Africa into a desert. The great alluvial plain of the Valley of the Po, in the north, was strong enough to withstand it. Ox-power, slave-power and the insatiable demand for the *panem* part of the *panem et circenses* that the Roman city plebs had to have as the price for keeping them quiet were, all combined, not enough to destroy that deep and resilient soil. For the Po Valley, like the Nile Valley, or the Gangetic Valley in India, or the Valley of the Yangtze Kiang, is the product of the mountains, and the soil there was, until recently, continually being renewed. Now, though river-training works confine to a large extent the waters of the rivers and prevent floods which were a nuisance to man, the annual renewal of silts from the high Alps is no longer applied.

But the Po Valley still grows and exports a great surplus of food, and this is achieved with a smaller and smaller labour force. Huge tractors on huge farms now tear into the sullen, heavy clay that was once overlain with fine silt, in so many areas now gone. Artificial fertilisers supply the elements that the silt used to supply and which the clay does not. Over two million tons of artificial fertiliser are applied to the soil of Italy every year: a great part of them in the Valley of the Po. For this great, flat plain lends itself to agribusiness in the way that the Oklahoma prairies did. The massive red-brick, red-tiled farmsteads built around courtyards testify to the high farming of Lombardy, when fine white cattle were reared and fattened in yards and their dung carted out in ox-carts to improve the heart of the land. Now most of the yards are without cattle, and the most cursory examination of the soil of the fields shows that it, too, is without something: it is without humus. Humus is the decaying remains of organic matter: the essential cement of the soil. You can grow large crops for decades in soil that has no humus – provided that is that you supply missing nutrients 'from the bag' as farmers say, or with artificial fertilisers. But under this regime your land will inevitably deteriorate and ultimately you – or more likely your descendants – will be able to grow practically nothing at all.

Hereditary farmers, who come from many generations of true agriculturalists, are stubborn about yielding to what are almost overwhelming economic drives of our time and they hang on, as long as they can, to sound agricultural practices. Such a one is Mr Sacchi who, with his brother, farms 370 acres, about 150 hectares, near Crema in Lombardy. We met him and went round his farm and filmed his operations. Our Mr Sacchi, a tough and vigorous man in his fifties, with some small children and two grown-up children (a son is a veterinary student and a daughter is an accountant), has the open-handed generosity and the hospitable instincts of the true husbandman.

The farm buildings were huge even by English East Anglian standards and the courtyard enclosed by them would have made a smallholding in itself. The dwelling house, on the north side of the quadrangle, had been an ancient monastic building in its time and was very beautiful. The signora was charming as only Italian ladies can be and the smaller children were friendly and fun. The interior of the house was luxurious. Certainly, whatever the farming was like outside, it worked.

Mr Sacchi had never heard of the organic movement in farming: he simply farmed the way he did because his training and his instincts told him that it was right. When so many of his neighbours had gone over to stockless agribusiness and monoculture, he had clung on to his livestock and rotation of crops. The farm carries a milking herd of 150 cows and a beef herd of 300 single-suckling cows and their calves. He fattens some of the latter and sells some as stores. The cows are outside for part of the time, on twenty-seven acres (eleven hectares) of permanent grass, but most of their time they spend in open or covered yards. Even the open yards have covered shelter. When under cover the cattle are on slats and their dung falls into lagoons below, is mixed with water, and pumped on to the arable land in the autumn. In the spring that land is rotovated. Mr Sacchi would far prefer, he told me, to keep all his cattle on straw and make properly composted straw manure but this would require far too much labour. Eighty-eight per cent of Italian people now live in the cities and therefore there are not enough people to work farms as they should be worked. But Mr Sacchi does make some straw-composted manure to put on his permanent pasture, for that he feels is important.

To add to the manure-producers there are sixty sows and their litters. With all this animal manure, the Sacchis do not have to use anything like as much artificial fertiliser as their neighbours: on the wheat land 130 lbs per acre (approximately 150 kilos per hectare) of nitro-chalk is all the nitrogen needed.

From thirty to thirty-five acres (twelve to fifteen hectares) of wheat are grown each year and twenty acres of barley. All this is drilled in the autumn. There are thirty-five acres of maize for grain and seventy acres of fodder maize. Surprisingly in these days thirty acres are down to annual rye-grass which is cut for hay or silage, and this ley ploughed up after one

Above, *As old as any city in the world, Warka now lies in ruins, surrounded by hundreds of miles of barren soil;* left, *The Roman ruins of Timgad in Tunisia where exhaustion of the soil left the city without food supplies;* overleaf, *Sheep grazing below date palms in Qurna, southern Iraq, one of the traditional sites of the Garden of Eden*

65

Above, *A flock of sheep in Kenya;* below, *Mycenae in Greece where olive trees survive on land with no topsoil. Uncontrolled grazing by sheep means it is unlikely the soil will recover*

Calabria, southern Italy: Above, Pentatillo was built on a hill to conserve good land and for defensive purposes; below, Well-tended land with intact terraces conserving soil

Vistagi in Crete and its inhabitants: Opposite, *Distilling wine to make* raki; above, *The village;* left, *A local shepherd*

71

Vistagi is almost completely self-sufficient and its inhabitants follow a traditional way of life

season. A variety of miscellaneous crops are grown on the rest of the acreage.

The maize for grain, which always follows the rye-grass ley, has increased 30 per cent in yield in the last thirty years, mostly due to fertilisers. Mr Sacchi told me that the constantly increasing cost of these latter though made him glad he had stuck to mixed farming. If nitro-chalk and urea (which he uses on his maize) were priced out of the market he could still grow quite good crops with the dung from his animals alone.

The milk from the milk cows is turned, on the farm, into cheese. This is a practice that is very much in accord with good organic theory, for the whey from the cheese-making goes to the pigs and the dung from the latter back to the land. Only the cheese itself, a high-value product, leaves the farm and fails to add to its fertility.

We filmed Mr Sacchi cutting his green maize with a forage harvester and it was a magnificent crop. Everything around him was healthy – his family, his pigs and his cattle. For they are all the products of a healthy *soil*. It was a sensuous pleasure, for me, to pick up a handful of the topsoil on that farm. The contrast between it and the stubborn yellow clay, baked to near-brick by the sun, of many of his neighbours was immense. But the labour force on this farm, including the two Sacchis, is eight men: far greater than it is on many a neighbouring agribusiness of the same acreage. The Sacchis might make more money if they sacked the men, got rid of the cattle and pigs, and went into agribusiness monoculture. That is, in the short term – in their lifetime. The price of oil goes up and down – according to the decisions of members of OPEC but in the long run it can only tend upward. The oil is finite – this world even is finite – and when the oil becomes scarcer its price will rise and then maybe the Sacchis' children will be glad that their father stuck to mixed farming and good husbandry. For the price of fertilisers and pesticides is tied directly to the price of oil and natural gas. The Sacchis make a moderate use of artificial fertilisers now but, due to their heavy animal stocking and mixed cropping, their land is in good enough heart to go on producing crops, *ad infinitum*, without them.

I then travelled south to visit some friends of mine who live in Tuscany. Their small farm is near Bibbiena, in a fold of the Apennines and is typical of smallholdings of the area. I also visited one of their neighbours, who has a somewhat larger farm which is more typical of the area.

This is hilly broken country and so was saved from the fate of the better land in Roman times, which was exploited by the latifundia – and thereby ruined. It did not pay big capitalists then, as it does now, to take over broken mountain country for their cash-crop farming. So since Roman times this hilly country has been in the hands of peasants, who have farmed it primarily on a subsistence basis and, it being their *own* land, and they completely dependent on it, they have looked after it.

The neighbouring farmer we went to see lives with his extended family in a large stone farmhouse, half-way up a steep, terraced hillside, and now farms some fifty acres (about twenty hectares). Thirty years ago, he had ten but the flight of the poorer peasants to the cities released more land and some of this he has been able to buy.

His land is all on steep hillsides and all is terraced. The terraces are level strips – a typical one being maybe fifty yards long and ten wide. The soil in them is retained by well-built stone walls. When I ask him when the terraces were built he just says, 'Roman times.' He then points over to the other side of the valley where there are more terraces, but these terraces have tumbled back to grass and weeds and – worse – many of them are broken. They have been abandoned. The lure of the cities after the Second World War took most of the *contadini*, or small peasants, from the land and many small farms and peasant holdings were simply abandoned.

Now the peasants who had the sense to stay on, such as our host, were able to take on some of this abandoned land but not all of it, because you cannot farm such land without human labour and the humans had mostly left. Our host pointed to terrace after terrace that had broken on the further slope. The stone walls had fallen at some point and the soil had washed through the gaps. The soil had gone for ever, it will never return. It had gone down to the stream, down to the river and down to the sea – soil that had been nurtured, cared for and improved since ancient Roman times had gone for ever in a decade. Our friend pointed to one abandoned farm and told us the sad story of how the man who owned it had left, twenty years ago, to work in a car factory in Torino He thought he would get rich but he didn't. He became bored and disillusioned by the factory life and had one day made the decision to take his wife and family home. He came back, and looked at his ruined, roofless house and his eroded fields – there was no soil left. Land that had given a living to hundreds of generations of mankind would give a living no more. *He* had betrayed it. He went back to the city for there was nothing else he could do.

On a happier note, our host took us round his own farm. He still worked a team of oxen, or at least his father did. But the Italians have been very clever at inventing small machines to take the hard slog out of farming on steep hillsides and on small acreages. One of these is the Olympia cutter, which is a cutter-bar machine, with a small petrol-engine mounted at one end of the cutter-bar. The operator walks behind it, controlling it with a pair of handles. I have seen one working, among trees, on a one-in-two slope. It cuts grass, rough scrub or herbage, and has an attachment for binding sheaves for when it is used for cutting wheat or barley. Our friend also had an excellent rotovator. He grew wheat, barley, oats and a variety of leguminous crops such as lupins and lucerne (alfalfa) and various clovers, for even the ancient Romans knew that the members of the *leguminosae* family increase the fertility of the soil. There were plenty of trees on his terraces: lemons, oranges, figs, some olives although this was high up and far from the sea, almonds and other fruits and nuts. A decent part of the acreage was down to grapes and the wine and brandy made from these, which we sampled, were superb.

A meal with these people is like a time-trip back to a more generous age. One receives an impression of *plenty*. Not plenty of money – but plenty of the very best food in the world, the best wine, superb olive oil, a fine, spacious house, good, solid, old-fashioned furniture and good, strong, serviceable clothes. The soil community of which these people are part is lucky in that the hilltops are covered with oak and alder forest and the terraced hillsides are well maintained. The forests were all clear-felled during the Second World War but have regenerated themselves since and are being carefully protected by the government. But it is sad to look over the other side of the valley and see the gaps in the terrace walls there and realise that the ancient soil – the necessary basis of all civilisation – is being washed away.

We travelled on south in Italy. The nearer we got to Rome the more eroded the country seemed to be and we passed many completely deserted villages and many others which had obviously lost much population to the big cities. We travelled to the very south of the country and centred ourselves at the modern seaside resort of Metaponto.

Metaponto itself, and the almost deserted motorway that connects it to the north, are examples of a desperate attempt to bring people and prosperity back to an eroded land. It was immediately obvious that the attempt had failed.

In some level areas in the valleys, fairly large-scale mechanised farming went on. The valleys had benefited from the loss of the hills. What soil had been on the latter had washed on to the lower ground and a fraction of it had lodged there. Most of the soil of Italy has ended up in the Mediterranean Sea.

We had gone there to film soil erosion for our television series. We could scarcely have chosen a better place. We could stop the cars anywhere, on any road, and simply set the camera up and turn it on and we would be filming soil erosion: unbelievable gullying wherever you looked. This land was but the skeleton of an ancient living soil community. It is easy to blame 'the Romans' but informed people to whom I talked said that soil abuse had been going on ever since the collapse of the Western Empire and, where it can, is going on now.

Under the present regime, the land of southern Italy, considered as a habitat for people, animals or plants, is doomed. If any little tree or shrub tries to grow, a goat or a sheep will come along and chop it. The climate, like the climate of all hot, dry, treeless climates, is inimical to life. There are no trees to absorb, retain and ultimately transpire the rain and so it rushes down into the sea taking the landscape with it. The cooling and moistening effect of tree transpiration is absent.

There are pockets of resistance. There are still villages with people in them, still little pockets here and there of soil retained in terraces and jealously guarded against sheep and goats – but the *heart* which has gone out of the land seems to have gone out of the people too. I felt that the latter were only waiting their chance to take off to the industrial cities of the north as most of their former friends had done before them. The vaunted motorway, built to bring people and prosperity *back* to the south, was only helping both to flee northwards to the fertile (and heavily industrialised) plain of Lombardy.

The ancient Greeks colonised southern Italy, and no doubt it was their great demand for timber for ship-building, pottery-firing and metal-smelting that struck the first blow at the forests that they found there. Sheep and goats made sure that the forests should not regenerate themselves as they do today. Cash-crop farming under the Romans began the destruction of the soil and this has been going on ever since. The task is accomplished now – it is the very rock that is being eroded at present for most of even the subsoil has gone.

The damage is not irreversible but only if there were a drive powered by a true religious fervour of the people could anything effective be done now. The Yeoman method of 'key-line' ploughing – chisel-ploughing across the slopes at a very slight angle to the contour lines, so as to lead surface water from the gulleys out to the ridges of the spurs – could green the eroded hillsides again. Sheep and goats would have to be completely banished, for a time at least, no matter what loss this would cause to individuals. Re-afforestation on a massive scale would be needed. No ploughing of sloping land without terracing would be allowed.

No doubt, if Italy ever awakes from the industrial nightmare which had led its people so astray, such a gigantic effort might be made. Hesiod wrote his *Works and Days* in Greece at the time when his countrymen were colonising the country round Metaponto and indeed planting colonies all round the Mediterranean world. In it he described the fate of men who care for their soil:

> Famine and blight do not beset the just,
> Who till their well-worked fields and feast. The earth
> Supports them lavishly; and on the hills

> The oak bears acorns for them at the top
> And honey-bees below; their woolly sheep
> Bear heavy fleeces, and their wives bear sons
> Just like their fathers. Since they always thrive,
> They have no need to go on ships, because
> The plenty-bringing land gives them her fruit.

But he goes on to describe what happens to men who look upon the land, not as a sacred trust, but as something simply to make money out of:

> But there are some who till the fields of pride
> And work at evil deeds; Zeus marks them out,
> And often, all the city suffers for
> Their wicked schemes, and on these men, from heaven
> The son of Kronos sends great punishments,
> Both plague and famine, and the people die.

*From the edition translated by Dorothea Wender, Penguin Classics, 1973. Reprinted by permission of Penguin Books Ltd.

Land and People in the Middle Ages
Herbert Girardet

With the decline of the Roman Empire in the fifth century AD, Europe was plunged into a state of uncertainty. The Romans had been harsh, and sometimes arrogant rulers, but they had brought a measure of stability to much of western Europe. Now that Rome's power was waning, the floodgates of invasion and conquest were opened. Many tribes had had to measure their strength against the military power of Rome. They had all learned something about the disciplined conduct of war, even in the misery of defeat.

Among the challengers were the Goths, the Vandals and the Huns, all of whom derived their military power from the fact that they were warriors on horseback. The Franks turned out to be the dominant force in Europe after the decline of Rome's power and were not challenged significantly as a dominant force in France until the invasion of the Arab Muslims in 732. They had started their territorial expansion from north Africa into Spain after the death of the prophet Mohammed. The Franks defeated the Arabs at Poitiers in France and managed to push them back across the Pyrenees. Their leader, and later their king, was Charles Martel, whose grandson, Charles – 'Charlemagne', Charles the Great – inherited the throne in AD 768. In the year 800 Charlemagne was crowned by Pope Leo as Roman Emperor of the West, the first person to hold that title for 300 years.

Under the Carolingians – the dynasty started by Charles Martel – revolutionary changes in the relationship between man and land were set in motion. This era witnessed the birth of feudalism, a system of land tenure in which the right to land was granted in exchange for military service. The king or emperor was the owner of all land. He gave his close associates the right over large tracts of it. In theory, these were tenures for life though in reality they became hereditary. The tenants or vassals usually divided their lands among sub-vassals. The vassals and sub-vassals farmed the land and in return owed their immediate overlords obedience, war service and a certain number of soldiers.

Land was still the primary source of wealth as it had been in Roman days. But now the productivity of the land and the peasants who worked it became a direct source of military power. While the vassals were obliged to provide foot-soldiers for the king from among the peasantry on their estates, they themselves were sworn to serve him as mounted knights. The horses were, of course, provided by the knights themselves and they were fed on grass and oats grown on their own estates, which had been granted to them by the king. Thus the military power of the king was directly identified with the productivity of the lands which he had granted to his vassals. Any vassal unwilling or unable to meet his military obligations forfeited the tenancy to his land. As Lynn White wrote in *Medieval Technology and Social Change,* the consequences were far-reaching:

The vassal class created by the military mutation of the eighth century became for generations the ruling element of European society, but through all subsequent chaos, and despite abuses, it never lost completely its sense of *noblesse oblige* even when a new rival class of burghers revived the Roman notion of the unconditional and socially irresponsible possession of property.

The massive territorial expansion of the Franks under Charles Martel and Charlemagne – they conquered much of France, Bavaria, Saxony, Lombardy, Venetia and other parts of northern Italy – was made possible above all else because of the highly effective cavalry supplied by their vassals. The Franks were the first European power to equip their knights with stirrups. This gave them a much greater control over their horses than mounted warriors had had up until then. Stirrups transformed the mounted warrior into a fearsome fighter. Being virtually merged with his mount, the knight's charge against his adversary was vastly amplified by the muscle-power of his horse. Stirrups stabilised the charge of the lancer and also the furious sweep of the knight brandishing his longsword. They enabled mounted archers to shoot their arrows with much greater power and accuracy.

Charlemagne made highly effective use of his knights thus equipped in his campaigns to consolidate his power in Europe. It was his grandfather, Charles Martel, who had first recognised the potential of the stirrup as a tool in warfare. But Charlemagne changed the system of land tenure to enable him to make full use of the greatly increased possibilities inherent in 'horsepower'. Since his vassals were required to be his knights and to provide additional cavalry troops from the estates they controlled, much more land than before had to be made available for grazing horses and for growing feed for them. Forests were cleared to provide new grazing land and more intensive systems of agriculture were introduced.

The process of forest clearance for the creation of farmland also had the benefit of providing timber for building and charcoal for smelting iron ore. During the age of Charlemagne much wider use was made of iron, for military equipment, for axes and also for farm implements. Water mills were built to grind corn but also to power bellows used in blast furnaces.

The more effective use of horses in warfare was soon to be complemented by their introduction into agriculture. Oxen had been used for pulling ploughs but the widespread application of a new collar harness greatly increased the pulling power of the horse. Heavy ploughs were introduced instead of the traditional ox-drawn scratch ploughs. These were much more suitable for the deep soils of northern Europe. When horses were also fitted with iron horseshoes they became highly effective in agricultural work like ploughing, harrowing and pulling carts. Thus horses gained key importance not only in warfare but also in food production. The feudal system took full advantage of these new possibilities.

Horsepower, waterpower and the controlled use of fire in blast furnaces added up to a formidable array of forces in the hands of medieval northern Europeans. By the twelfth century AD horses had taken over as the main traction animals on their farms. More land per farmer could be taken into cultivation as horsepower greatly amplified human muscle-power.

The technological innovations of the Frankians did not remain their exclusive property. They quickly spread throughout Europe and beyond. (In fact, one must emphasise that medieval technology was a fusion of practical ideas which had their origin in many separate places; for example, the prototype of the stirrup appears to have originated in China.) The cavalry army of the Normans, who invaded Britain in 1066 from their base in northern France, made use of the superior continental war technologies against whom the more traditionally equipped Anglo-Saxon foot-soldiers were unevenly matched. William the Conqueror, with the help of his knights in armour, took over as King of England. He was crowned in London on Christmas Day, 1066.

The new king then proceeded to divide the soils. His principle was a simple one: he owned all the soil in England, and allotted certain portions of it to his deserving followers as tenants; if they had no lineal heirs, the land would revert to the crown. England was the perfect feudal state. The holders of the great fiefs were tenants-in-chief or barons. They provided for their retainers by subinfeudation, granting them land with the usual obligations of vassals. Thus there were about five thousand knights at the king's call. To enforce his rule, William decreed the immediate construction of royal castles throughout the kingdom. (M.Bishop, *The Pelican Book of the Middle Ages*.)

Anglo-Saxon villages, which had been enjoying a fair degree of self-determination, were now overshadowed by Norman castles. Landowner-ship was converted into tenancy. The villeins became sharecroppers in the service of their new feudal landlords. The king and his knights annexed all the forests in England for their own use. This was a great blow for the peasants since the forests and woodlands had represented a major resource for them: they had provided firewood, grazing land and hunting grounds. From now on firewood had to be bought; for a wood-penny the villagers were now entitled to what wood they could manage to take out 'by hook or by crook'.

From their newly-built castles the Normans enjoyed hunting expeditions in the forests of which they had taken sole possession. Any peasants found 'poaching' there were dealt with severely. The Normans set about clearing the forest land for farming purposes and also for the production of charcoal which they required for their new iron industries.

When William was firmly established as King of England he commissioned a register of properties, the famous Domesday Book. This proved to be a most valuable tool for his administration, particularly for the purposes of taxation. The size and quality of land holdings are registered with considerable accuracy. Human and animal population of farms and villages are recorded. Every water mill in the country is noted: 5624 of them in all. This is an astonishingly large number, considering that the population of England was only around 1.4 million.

Among many other important innovations the Normans brought with them was the three-field rotation which had a major impact on farming in Britain, as it did in continental Europe where it had been popularised by Charlemagne. Under the three-field plan the arable land was divided roughly into thirds. Winter wheat or rye was planted in one field, oats, barley, peas, chick-peas, lentils or broad beans were planted in the second. The third field was left fallow. Each year, the crops were rotated to leave one field fallow and ensuring that the same crop was not grown on the same field in successive years.

The three-field system greatly increased the productivity of the land. By introducing legumes into the rotation, the supply of nitrogen to the soil was more readily assured, enabling the farmer to shorten the fallow period. The fact that he could now grow oats as part of his farming cycle made it easier for him to introduce horses as draught animals instead of oxen. Horses – when fed on grass as well as on oats – are faster and stronger than oxen, capable of pulling the heavy ploughs necessary to turn over the deep soils of northern Europe. Horses greatly increased the productivity of European farmers, enabling them to produce the surpluses demanded by their feudal overlords.

During the twelfth and thirteenth centuries a rapid increase in agricultural production in north-western Europe coincided with a dramatic population increase. The rural population doubled between the years 1100 and 1300, and so much farmland was reclaimed from forests, swamps and hillsides. At the same time, labour productivity also more than doubled, with the introduction of horses, watermills and then windmills. Rural prosperity was undermined by the heavy demands of feudal lords on the rural population. Up to 50 per cent of the harvest had to be handed over to them. In many parts of Europe the peasants were not able to enjoy a living standard much above subsistence level.

Kings and feudal lords made much use of the huge potential inherent in horse-, water- and windpower. They used these newly-harnessed natural forces to strengthen their economic and military power. The Church, too, took a stake in the new productive forces, particularly through the pioneering efforts of the Cistercian Order. In the twelfth century the Cistercians had 742 monasteries in France alone, all of which owned extensive land holdings, usually reclaimed from swamps and forest land, which were farmed by the monks according to the latest techniques. The monks themselves took a hand in farming, guiding the lay brothers – who did much of the agricultural work – in the practice of three-field rotation, the use of horses and wheeled ploughs, the skills of shepherding and viticulture, and the effective use of watermills. The Cistercians built watermills in every one of their monasteries, they were used to crush wheat, sieve flour, full cloth and tan leather with machinery attached to them. They probably also harnessed waterpower to activate bellows in some of their smithies and breweries. The monasteries were equipped with piped water and sewers for cleanliness and hygiene.

The scientific approach of the Cistercians to land clearance, farming and the use of mechanical power made them a highly effective force in the medieval economy. In the south, they became major producers of wine which they sold to countries such as Holland and Britain. In the north, and particularly in Britain, they became very successful at producing wool for export. Every monastery aimed at being largely self-sufficient in food, including wheat, vegetables, fruit and meat. Beer and wine, which were produced by the monks themselves, were an important part of the monastic diet.

The Cistercians were the main Christian pioneering order. They carried out the Old Testament command that man should have 'dominion over the fish of the sea, and over the fowl of the air, and over the cattle, and over all the earth and over every creeping thing that creepeth on the Earth'. The monks themselves had to refrain from obeying the call to 'be fruitful and multiply', but they did 'replenish the earth and subdue it.'

'Christianity did encourage . . . the transformation of wildernesses, those dreaded haunts of demons, the ancient nature-gods, into farm and pasture. But in general terms, what men had to transform was not nature, but themselves, and even that was possible only with the aid of God's grace.' (J.Passmore, *Man's Responsibility for Nature*.)

The frugality of the monks and their disinclination to accumulate possessions and wealth enabled the monasteries to invest their considerable profits from trade and rents in the acquisition of more and more land and productive equipment for farms and workshops. The Cistercians played an important part in spreading metal technology throughout Europe. They were running the most modern factories in Europe:

All over Europe the Cistercians set up around their monasteries a whole range of granges and model farms . . . For example, for the monastery at Les Dunes, the lay brothers converted some 25,000 acres of the sandy and marshy deserts of Flanders coastline into fertile soil, while for the monastery at Chiaraville, near Milan, they built an irrigation canal, completed in 1138, to bring water to the crops. (J.Gimpel, *The Medieval Machine.*)

By the beginning of the fourteenth century economic and population growth in northern Europe had transformed the face of the countryside. Millions of acres of forest land had been cleared and turned into grazing and farmland. Almost everywhere castles had been built by the new 'horse-powered' lords of the land. New towns sprang up in many places which were filled with people from farms and villages. They left the land because the introduction of the horse and new farming technologies rendered them jobless; because of rapid rural population growth; because of the heavy demands made on them by feudal landlords; and because of the new opportunities offered to them in the towns. A great attraction for poor serfs from the country was that they were entitled to become free men within the confines of the town walls after a year's residence there. The emergence of craft industries provided many newcomers with work. The manufacture of woollens, linen, armour, furniture, wooden carts, tools, weapons, glass, leather goods and many other products was concentrated in towns.

Dinkelsbühl in Franconia, which today is a part of Bavaria, is a medieval town which has survived virtually unscathed until today. One of the very few towns in Europe whose ring wall is still there for all to see, Dinkelsbühl today displays its flawless medieval appearance to many visitors. It was bypassed by the industrial revolutions of our age and is still a country town with cobblestone streets. The current street names give an indication of the kinds of crafts that made Dinkelsbühl prosper, notably metal-working and cloth-making. The wine market near the centre of the town played an important part in trade with distant cities such as Cologne and Amsterdam. Dinkelsbühl was well-located on one of the major trade routes from southern to northern Germany and Holland.

Craft production was controlled by the crafts' guilds which had a strictly controlled membership. It did not take long before they became set in their ways, unwilling to give newcomers or new production methods a chance. The cloth-makers and blacksmiths of towns like Dinkelsbühl were keen to keep the money they made in their own pockets. When they could afford it they built large and beautiful houses to live and work in. These are still there for us to admire. And so is the imposing Gothic church which they helped to finance.

Other centres of commercial activity were the leather market and the corn exchange. A brush-making workshop and a timber market could also be found in the town. They all depended for their raw materials on the fertility of the countryside around Dinkelsbühl. In fact, much of the land outside the town belonged to farmers who had their farm buildings inside the town walls. Today there are still a few such farms near the Nördlinger Tor. Cows are still milked every morning and every night within the city walls. Their fodder is obtained from fields outside town and their dung – composted with straw – is conscientiously returned to the land. In medieval days, human waste, too, was returned to the fields from the town by people especially employed to do this job. A town like Dinkelsbühl – which today has 11,000 inhabitants – is not too big to return its human waste to the surrounding countryside. Huge cities like Rome in its heyday found it impossible to return the fertility taken from the land back to where it came from. Small towns, on the other hand, that had grown out of the local countryside, could retain a symbiotic relationship with it.

By the middle of the fourteenth century the 'four horsemen of the Apocalypse' were charging across Europe. Famine, war, pestilence and death held the continent in its grip. Trading ships brought the plague into Italian harbours from the East in 1347 and it spread across Europe carried by rats with plague-carrying fleas. The disease took a huge toll on Europe's population: between 30 and 50 per cent died in just a few years. H.A. Walters, in his book *Ecology, Food and Civilisation*, discusses the effects:

The net result was that in the year 1400 the total population of Europe was again reduced to about forty-five million, or two-thirds of what it was at the peak of Roman power in AD 200. Many cultivated areas went back to the wild state, reclamation ceased and, broadly speaking, the ecology of the countryside underwent a great change where the hand of man was lifted from it. Also various wars contributed their toll causing populations to flee the countryside.

The ravages of pestilence, famine and war led to the desertion of many thousands of villages throughout Europe. Farms and villages in remote places and on difficult land were abandoned and the people from these places migrated to 'better pastures', to more fertile regions where the population had been drastically reduced. New outbreaks of the plague took a further toll of these settlements. Hundreds of thousands of farms lost their tenants and reverted to scrub and woodland. In many places landowners found it impossible to get people to farm the land.

After the plague subsided towards the end of the fourteenth century, the relationship between landlords and tenants became very brittle throughout Europe. The price for farm produce dropped everywhere, yet tenants still had to pay high rents to the lords. Peasant revolts broke out, first in England in 1381. There was a shortage of labour as a result of the plague but the landlords refused to increase wages or reduce the feudal burdens of the serfs. The rebellion was led by Wat Tyler and a priest, John Ball, who preached love and peace as well as equality:

Ah, ye good people, the matter goeth not well to pass in England, nor shall not do so till everything be common, and that there be no villeins or gentlemen, but that we may be all united and that the lords be no greater masters than we be. What have we deserved or why should we be thus kept in serfdom? We be all come from one father and one mother, Adam and Eve. (M.Bishop, *The Pelican Book of the Middle Ages.*)

The rebellion was put down by force of arms. In continental Europe, too, peasant rebellions broke out in the following decades. In Germany, peasants protested against oppression in the 1430s. They had very little success against the lords. Nearly 100 years later the Peasant War broke out all over southern Germany. Partly stimulated by Martin Luther's teachings on Christian freedom and responsibility, the peasants demanded justice before God, and tried to throw off the bonds of serfdom. In 1524 they published their Twelve Articles, the third of which stated their case clearly:

It has been the custom hitherto for men to hold us as their property, which is pitiable enough considering that Christ has delivered us and redeemed us all by shedding his precious blood, the lowly as well as the great. Hence it is consistent with the scripture that we should be free ... We therefore take it for granted that you will free us from serfdom as true Christians, unless it be shown from the Gospels that we are serfs. (W.D.Camp, *Roots of Western Civilisation.*)

The badly-armed peasants were no match for the combined forces of the German nobility. They found it difficult to organise into effective military units, scattered as they were on small farmsteads, tied to the seasonal requirements of farm life. Only in the principality of Salzburg – then part of Germany – did the peasants, reinforced by miners and the burghers of the city, succeed in putting up a fight whose outcome was uncertain. In 1525 peasant armies under the leadership of Michael Gaismair had several important victories against the forces of the prince archbishop of Salzburg, Cardinal Matthaeus Lang. The rich and powerful principality of Salzburg was one of the few territories in western Europe which was under the control of the clergy. Under Lang, the rule of the church became extremely oppressive, uniting much of the population in organised resistance.

The anger of the peasants was directed against the Cardinal as the principal landowner. Virtually all the land, in the fertile plains as well as in the mountain valleys of the Pinzgau and the Pongau, was owned by the prince-bishop. The social and economic position of those among the peasantry who were in charge of a farm was fairly acceptable. However, many serfs, farmhands, milkmaids, as well as poor traders and craftspeople were virtually destitute. There were also numerous farmers' sons who were economically insecure because only the eldest son was legally entitled to take over a farm from his father. Younger sons of peasant farmers constituted a reservoir of discontent and rebelliousness which played an important role in the peasant wars.

Their anger having reached breaking point in 1525, they rose in rebellion against the Archbishop. It was the night of June 5. Drums beating, mountain fires blazing and bells ringing, the peasants approached from the different areas of his domain – the Pongau, Gastein, Radstadt – entering the city through the Stein Gate. Lang fled to the fortress while the peasants invaded the Residenz, destroying its possessions and banqueting on its food. With wooden canons they answered the thunderous artillery emanating from the Festung. (D.Burgwyn, *Salzburg, a Portrait.*)

The rebel army besieged the mighty castle of Hohensalzburg, still towering above the city today in its granite rock, for three months. They organised their own system of government, food supply, taxation of wealth, mining activities, courts and so on. The profits from the Hallein salt mines were used to pay the soldiers. But, in the end, the peasant forces were defeated by Ludwig, Prince of Bavaria, who sent a well-equipped army to 'restore order'. The peasants retreated in despair. Those who were captured were thrown into dungeons. When another rebellion broke out in the following year, Lang had his mercenaries capture and murder the peasants by the thousands. The feudal powers had the upper hand once again.

The sixteenth century was a period of dramatic population growth throughout much of Europe. Who knows how the confrontations between peasants and landlords over landownership and rents would have been resolved had not the Europeans colonised America. Not only did it yield land for European settlers – to the detriment of the indigenous people – but it also became the source of a vast amount of gold and silver. The King of Spain was the ruler who benefited most from the riches of the Aztecs and the Incas which were brought to him by the conquistadores. He spent vast sums on his campaign to defend Catholicism against the Protestants in Europe and in the effort of Christian colonisation in America. Between 1503 and 1650 approximately 185 tons of gold and 16,000 tons of silver were transferred from America to Europe. This was used to stabilise currencies and to stimulate manufacturing production for the colonisation effort. There can be no doubt that the gold and silver of the Aztecs and the Incas, taken by the Spaniards, helped ultimately to finance Europe's industrial revolution. The Spanish themselves did not encourage industries in their country. It was Holland, Flanders, France, Germany and Britain who benefited from the stimulation of manufacturing.

In the sixteenth and seventeenth centuries, population growth, the expansion of manufacturing industries, increased demand for timber for house construction and shipbuilding led to renewed pressure on forests. In fact, as early as 1492 an Italian chronicler reported a great shortage of timber. It was impossible to find oak timber any more and people had started to cut down domestic trees. Even olive groves were being destroyed in the search for timber and firewood. The destruction of forests was to happen throughout Europe. Wherever mines were opened, there was a great demand for pit props. Charcoal for smelting metals also took a huge toll on the trees of France, Britain and Germany.

The rapidly increasing price of timber and charcoal led to the curtailment of mining activity in some areas. In Britain the armaments industry was severely affected by a shortage of charcoal. From 1632 onwards she started to import iron canons as well as timber from forest-rich Sweden. Britain turned to coal for domestic heating as well as for metal-smelting. The introduction of the use of coke had a dramatic impact on iron and steel production. Britain with her rich coal deposits was the first country to make extensive use of this resource. The use of coal in industrial processes took the pressure off forests, first in Britain and then in the rest of Europe. But it also gave rise to air pollution and thus introduced a new kind of environmental impact. Once the controlled use of fire was extended to the generation of power in steam engines, this impact was further intensified. 'Fire-power' gave vastly more 'muscle' to style ourselves as masters over nature than horse-, water- or wind-power before it. And at the same time man changed from a largely soil-dependent creature to one who had to rely on stored soil productivity. The growing dependence on fossil fuels transformed man's relationship to nature more than any previous new adaptation. With the steam-engine and internal combustion engine at his command his powers became truly amplified

A Look at England
—— John Seymour ——

The first modern writers who really came to an understanding of the rationale of the medieval strip system were C.S. and C.S.Orwin who wrote a book about it, *The Open Fields*. Briefly, their explanation was this. The inhabitants of, say, an Anglo-Saxon village in England would decide to clear a piece of woodland to bring more land into cultivation. When the clearing was done each plough team would get to work, ploughing the land. Now I have ploughed with oxen, in South Africa, where we used to pull a three-furrow disc plough with ten oxen. The first thing you find when you start to plough with a long team of oxen is that they are very cumbersome to turn. You therefore try to arrange it so that you turn them as little as possible. So, the natural thing for the Anglo-Saxon ploughmen to do was to plough a long narrow strip and to go on ploughing until they had done a day's work. This would have constituted one medieval strip. Now they would have to return to the village with their oxen for the night and so it did not matter to them if some of their fellow villagers had gone out and started to plough more strips alongside their strip. They would simply move further along the field the next day and start ploughing another strip maybe several strips away from their first strip. Thus, at the end of the ploughing season, each householder, or plough-owner, ended up with a number of strips scattered right across the great open field.

Anybody who holds land where fences are non-existent or primitive does not have to be told how difficult it is to control domestic animals in such circumstances. The medieval peasants had to simplify their fencing arrangements as much as they could. To do this they divided the arable land up into three great fields and everybody had to turn to and fence two of these against livestock each year, with temporary fences. In one, winter wheat and rye was grown. In the other, spring-sown grain: mostly barley, peas and beans. The third field was left fallow and the animals of the village allowed to graze over it. Thus this third of the arable land was manured and given a rest.

The system had many advantages. The strip system was extremely flexible. Each man had as much land as he was able to plough with the

Opposite, *An Amerindian woman digging up a warifteine plant which is traditionally used for contraception; overleaf, Brazilian tropical rain forest before and after clearance for a cattle ranch. A few years later the soil will be exhausted and the land will change from infertile scrubland to desert*

Opposite above, *A cattle ranch in Rondônia, Brazil;* opposite below, *A small-scale maize plantation, Brazil, which will allow the forest to regenerate itself;* above and below, *Modern and less recent dust storms in Niger and America illustrate the havoc wreaked by intensive farming methods*

Above, *The devastating effects of erosion in Madagascar;* below, *A tree clings on to the rapidly-eroding soil in an area where forest is gradually becoming a desert*

Above, *Gulley erosion in Kenya – a gulley may double its size in a year;* below, *Sheet erosion in Idaho, America, has removed the soil down to the bedrock*

oxen at his disposal. If a man owned a plough and one ox, and needed six oxen to pull the plough, he went in with five ox-owning neighbours and they ploughed each man's strips in turn. A poor man, or a newcomer, might be awarded one or two strips, just to get started with, and would let these strips for money, or pay (perhaps with credit) someone to plough them for him, and thus be able to improve himself. A widow would be allotted some strips and would get neighbours to plough them for her – thus she would not be left destitute. People had to fit in with the general cropping plan, and thus experiment was precluded, but at least every man could manure his own soil, was responsible for weed control and harvested his own crop. It was not the fault of the strip system that he had to hand at least half of the latter over to this lord of the manor. The open-field system lasted a thousand years and under it the land did not deteriorate.

But as more people went to live in the cities and so a greater proportion of the crop could be sold for cash to them, and as farming methods were improved and changed, the open-field system began to come under pressure. The lords of the manor had always had their own enclosed land (their *demesne*) that was farmed for them by their bailiffs, using the free labour that the peasants were bound to give them for so many days a year. The lords began to experiment. They found that they could grow more crops on enclosed land than the peasants could on their strips. The strip-holders began to notice this.

And then there came a move, perhaps shortly after the Black Death, to enclose the land: to give each man a single block of land which he could surround by a ring-fence and farm exactly as he liked.

Acts of Parliament were passed to allow this to happen and, although the process took several centuries, eventually all the land of England was enclosed excepting some remnants of common land, and a little area at Laxton, in Nottinghamshire, which escaped the process, somehow, to this day.

Besides the arable land, much of the common land was enclosed. Some of this was done by Act of Parliament, much of it quite illegally. After the Reformation and the destruction of the church, there was no effective moral curb on the rapacity of the big landowners.

At the Reformation, too, all the land was taken away from the monasteries and handed out to Henry VIII's favourites. The latter, who tended to be more money-orientated and less inhibited by custom and the old tradition of *noblesse oblige* than their aristocratic predecessors, were quite ruthless in turning the old tenants of the monasteries out and letting their arable lands tumble down to grass to be grazed by sheep, which were then the most profitable crop.

Opposite, *Death Valley, California, where erosion has reached its ultimate conclusion*

The result was to drive thousands of people off the land altogether, many to roam the roads as vagrants and beggars, most eventually to swell the ever-increasing numbers in the growing cities and fuel the industrial revolution, and to turn what had been a non-extractive subsistence agriculture into an extractive money-rewarded one. Rich men became richer, poor men poorer and, at first at least, the manner of farming the land was improved – from the point of view, that is, of the land itself.

Barns, I should think, two hundred feet long; ricks of enormous size and most numerous; crops of wheat, five quarters to the acre, on the average; and a public house without either bacon or corn! The labourers' houses . . . beggarly in the extreme. The people dirty, poor-looking, ragged and particularly *dirty* . . . Invariably have I observed that the richer the soil, and the more destitute of woods, that is to say the more particularly a corn country, the more miserable the labourers. The cause is this, the great, the big bullfrog grasps all. In this beautiful island every inch of land is appropriated by the rich. No hedges, no ditches, no commons, no grassy lanes: a country divided into great farms; a few trees surround the great farmhouse. All the rest is bare of trees; and the wretched labourer has not a stick of wood, and no place for a pig or cow to graze or even lie down upon.

Thus did William Cobbett, writing in the first decade of the nineteenth century, describe the Isle of Thanet, a part of England as it appeared to him *after* the enclosure movement was completed. I drove across the Isle of Thanet in a governess cart, drawn by a pony, on a camping holiday that took me from Ramsgate to the border of Cornwall, and can report that the land has not changed since Cobbett's day, except that more of the land is down to cabbages and other brassica crops and less to wheat; but it is still treeless, hedgeless, has no commonages, and gives a strong impression of ugliness and comfortlessness. When Cobbett called it a beautiful island, I think he meant that the *soil* was beautiful. The soil is beautiful now – provided you lace it generously with nitrogen, potash and phosphate – otherwise I do not believe it is much good at all. My pony ride took me to a place that *was* very beautiful though: a farm run by the nuns of a small nunnery down by the River Stour. These holy ladies milked a small herd of Jersey cows, kept poultry, bees, cultivated wheat for their own mill and bakery, and kept a superb market garden. I saw a nun on a tractor which was pulling a trailer-load of beautiful cow manure, and I felt that she was the holiest nun I am ever likely to see.

The medieval agriculture described in the last chapter was a kind of low-key agriculture. It did not severely damage the soil – but it did not improve it much. It did not produce very high yields of wheat per acre, or per man-hour, but it fed Europe adequately, and provided enough surplus for trade to finance the building of Chartres Cathedral. At the time when Cobbett wrote about the Isle of Thanet, the old open-field system of the Middle Ages had been almost completely supplanted in England by a system of large, enclosed, tenanted farms, on huge estates, and the great bulk of the old peasantry had been displaced from the land and sent off to wander the roads as beggars, join the army or navy, or find work in the growing industrial towns and cities. The few men and women left on the land, working as landless labourers for the new breed of big farmers ('the big bullfrogs') were oppressed in the extreme. Their cottages were mere hovels, their wages were ridiculous and they were hardly allowed enough of the produce of their severe labour to keep their children alive. If they were caught snaring a rabbit they were shipped off to Australia.

But we are concerned in this book not with the treatment of people but with the treatment of the land. For no matter how well a social system treats people, if it does not treat the land well the people will ultimately suffer. The most highly paid labourers in the world will starve if the harvests fail. No trade union is going to save them.

The land of France fared wonderfully after the French Revolution, because for the first time for many centuries the cultivators owned the soil. If a man has a piece of land and knows that it is his, and that he can leave it to his children, he will tend it with loving care and devote his life and energies to the improvement of it. He may even plant trees on it. In England, after the enclosures, when the common land was shared out among the commoners for each to enclose his share within a ring-fence and clear it and plough it, after the great monastic houses had been pillaged and their lands given out to the king's favourites, and after the smaller peasants had been evicted by big landowners and their tenancies given to large farmers, a different kind of agriculture developed. Socially inequitable it might have been but, at first at least, it was marvellous for the soil.

The squires, or big landlords, planted trees about their estates. They did this for three reasons: one, they liked shooting and hunting and woods provided coverts for game and foxes; two, they were men of taste and they liked the look of trees and parklands, and three, they knew that they were going to leave their estates to their children and grandchildren and wished these to inherit maturing woodlands which would, in due course, make them rich. They supposed, too, that these heirs would plant more woodlands, to replace the ones they cut down.

The squires served two other good purposes. On their own home farms they experimented with new agricultural ideas. These ideas were mostly beneficial because the squires had no dangerous chemicals or powerful machines and they could do little harm to the soil with men and horses. The second good purpose was that they helped and encouraged their farmer tenants to improve their soil. The most famous of them all – Coke of Norfolk – would build, free, for any tenant, a cattle-fattening yard, provided the tenant agreed to keep it full of cattle, and he charged no increased rent for this amenity. The result was a rapid improvement of the hitherto light and hungry land over a large part of north Norfolk as great quantities of muck were carted out and spread on it. The same man greatly encouraged the folding of sheep on the arable land. This was the process of growing winter fodder crops, such as swedes or turnips, and penning sheep on the crops within small pens which were moved every day.

When I was boy I spent a long winter on a farm in the Cotswold Hills where this method of sheep husbandry was still employed. It was a labour-intensive job – I know because I was the labour. But, when I visited the farm a year later, I could see a sharp dividing line in the barley that occupied what had been the sheep-and-turnip field – with the barley much better one side of it than the other. The line represented the limit of the land folded off to sheep when I was there. When the sheep got to the line they were sold off fat and so the remainder of the field was not 'sheeped'. That was the poorer piece. Sheep, which wreak such terrible havoc to the soil when allowed to run too freely on unsuitable land, do marvellous good when thus folded over an arable crop: their hooves were called 'golden hooves' by the eighteenth- and nineteenth-century farmers.

That farm in the Cotswolds, on which I laboured in the early 1930s was one of the few good examples of the original English feudal system which remained in its final form. The estate belonged to Lord Sherborne, who lived in the ancestral seat near by. The farmer I worked for, who rented 350 acres (about 140 hectares) of light Cotswold land, was named Garne. Our next-door neighbour was an Abear. The Garnes and the Abears were henchmen of the first Lord Sherborne, and had sailed over to Hastings with him in 1066. Both families, and indeed the squire himself, were direct descendants from the Conqueror's soldiers in the male line.

Arthur Garne of Cocklebarrow Farm farmed in a manner most beneficial to the soil. Besides a small herd of dairy cows he had a herd of Garne Shorthorns. There were fifty breeding sows and their litters, all kept out on the land on free range, a flock of pure-bred Cotswold ewes, and 200 fattening 'tegs', or young sheep and six working horses. The manure from all these animals kept the land in excellent heart and, with very small inputs (except labour), the farm produced fine crops of wheat and barley.

One of the Abear sons was about my age, and he and I (after a twelve-hour day of gruelling manual labour which began at half past five in the morning) used to cycle, sometimes, into Cirencester, some thirty miles there and back, to go to the cinema. We thought nothing of either the work or the ride: both did us good. Once when we were doing this, the young Abear told me that his father was losing money on his folded sheep. I, from the lofty height of having spent three years at an agricultural college, said, 'Why keep sheep then? Arable sheep are old-fashioned.'

'Yes – but they *do* the land,' was his reply. This was a farmer's way of saying they did good to the land. No modern conventional farmer would let such a consideration enter his head. He knows that – whatever the state of his land – he can get such growth of crop as he needs by putting on chemical fertiliser. He thinks he does not need to 'do' the land. We shall know if he is right by the end of this century.

To see a typical existing large estate we went to the house of Lord Digby, at Minterne, in Dorset, southern England. Typical, we found, it was not. Lord Digby has strong views about trying to keep country people in the country, and therefore he farms his home acres in such a way as to retain, as far as possible, the fertility of the land, the beauty of the countryside, and the human population.

Before I arrived at the Digby estate I visited a neighbouring estate, of very similar type of country and size but run by a London businessman on conventional agribusiness lines. I saw about 1500 acres (about 600 hectares) from which practically every tree had been cut down as the only short-term way of making money out of trees is by cutting them down and not replanting them. The hedges had all been bulldozed (there had never been any ditches for this was chalk country and therefore naturally draining), there were no four-footed animals on the farm except some pet horses, the whole acreage was down to wheat and barley, all straw was burnt. This monoculture on a debilitated soil made a massive poison-spray programme essential, to keep pests within bounds. As for humans, except for three tractor drivers there were no farm workers, all the farm cottages had either been pulled down or sold to Londoners. It was obvious that the owner was making a great deal of money. By the time all the thin soil had eroded from his chalk hillsides he would have made enough money not to have to worry about farming and at least, if he had an heir, although he would have no decent land to leave him, he could will him plenty of stocks and shares.

Lord Digby, by contrast, farms 1600 acres (650 hectares) of which 200 acres are woodland, 300 rough grass, 300 good permanent pasture and 800 arable. A proportion of the arable land is always down to grass-and-clover ley. He has 180 milking cows, divided into two herds so that the manure from them can be more evenly spread on his rather long and narrow farm. When I was there he was in the process of taking on a shepherd and intended to build up a flock of 1700 ewes. This farming operation was run by nine full-time farm workers, with part-time workers too. The estate employed and supported fifteen families altogether: a total population of forty-four people. Lord Digby owned twenty-four houses and cottages, besides his own, half of them rented to outsiders, half of them to housing workers on the estate. Lady Digby is interested in crafts, and encourages craftsmen to take over the many empty outbuildings on the estate. There is talk of starting a craft centre in an old mill.

The great house, which is enormous, was built in 1906, after the seventeenth-century house was pulled down. It is far too big for the sole use of a single family, so half of it has been divided into seven flats which are let to people. The other half, which is occupied by the family, is made much use of by the people of Dorset. The Dorset Opera and a Summer Music Society give their performances there; the Dorset Community Council uses the house as its headquarters as do Dorset Youth Clubs, the Victoria Society, the Girl Guides, and the Boy Scouts. The Institute of Directors convenes here sometimes as does the Dorset Society for Under-Fives. The fine ornamental gardens are open to the public.

Obviously under this regime the countryside flourishes and so does the soil. In the present financial climate, though, Lord Digby's bank balance does not. If he did not also have city business interests he could not continue and would have to fell his trees, sack his men, sell off his cows and his sheep, and go in for grain monoculture like most of his neighbours.

I am not holding a brief for big estates. Most big estate owners in Britain have completely betrayed their trust to the land. It is interesting, and heartening, to find one who has not. On most big estates in England, before they were broken up by death duties, changing economies and two world wars, the cubic capacity of the lord's house far exceeded that of the rest of the housing on the estate combined, including the doctor's house and the parsonage. Surely this is an inequitable state of affairs? But the 'Big House' *has* a part to play, as a centre of social life and culture, if it is properly used. And when we think of the 'beautiful English countryside' we undoubtedly envisage something very like the Digby estate.

In contrast to this I must now describe what has come to be a much more typical way of treating our countryside. I have on my desk an article which appeared in a farming magazine. It extols in extravagant terms the farming methods used on an Essex farm. The farm (which I know) is on very deep alluvial soil, about half-silt and half-clay. It would grow excellent oak trees, walnuts or other high-quality hard woods and magnificent apples and other top-fruit. The 2500 acres (1000 hectares) of the farm could support twenty-five farmers and their families each with 100 acres or 250 smallholders with ten acres apiece. One could imagine two or three fair-sized villages supported by such a rural community, surrounded by woods and orchards and small farms: that land could be a paradise.

As it actually *is* we read from the magazine's account. The article is headed with a photograph of the farm. This shows a Land Rover at the end of a farm track which slants, dead straight, to the edge of the picture. For the rest – there is *nothing* but a great expanse of wheat and the sky. The impression that the picture leaves is of extreme ugliness – it depresses the spirit to look at it. The caption makes the point that this is no place to run out of fuel, for it is three miles from the farm buildings. The article reads:

This is what happens. Straw burning starts as soon as there is a safe space behind the combines at harvest. Then a 145-horsepower County 1454 goes in with a heavy Doe-type cultivator to pan-bust down to fourteen or fifteen inches. Working at a rate of thirty acres a day, this operation is often right up behind the combines. This is immediately followed by a 100-horsepower County 1004 pulling two offset sets of discs in tandem or, if the soil is too hard to disc, a Lantrac chisel cultivator. This stage of operation is usually completed by the end of September, leaving plenty of time for grass seeds to germinate before drilling. The last weeds are killed off in early October with paraquat at the full rate of one and a half pints an acre.

But this massive use of giant tractors powered by the world's dwindling supplies of fossil fuel is only a beginning. The real work on this farm is done by poisons. The polite word for poisons on the land is now biocide, it sounds less sinister but means the same thing.

It would be wrong to think that it's minimal cultivations alone that have solved a sticky farming problem and created a streamlined and profitable enterprise. As stated at the beginning of this article, this has only been achieved by a total commitment to chemicals as well.

The move to a five-year run of winter wheat brought with it all the expected weed and pest problems, and some more besides. Most noticeable was the build-up of wild oats and blackgrass. These two weeds are now being severely bashed by Dicurane in the autumn and by Suffix in the spring – both sprayed over the entire acreage for two years running. The paraquat sprayed in September must also get rid of some seedlings. This blanket spraying campaign has been expensive but successful.

The article then enumerates all the sprays used on this land every year, and they are: paraquat, Dicurane, a broad-leaved weed herbicide, Suffix, an aphicide (something to kill aphids) and Benlate. The article continues:

The enterprise is a superb example of how advanced techniques which are thoughtfully applied can make possible a cropping programme that would not have been possible even as recently as ten years ago.

This article was written in 1974. There are many more poisons, with even stranger names, in use today.

And what are we to suppose is the effect of this massive annual drenching of nearly all the arable land of Europe with deadly poisons decade after decade and, if it goes on, century after century? What effect is it likely to have on other forms of life besides man? What effect is it supposed to have on man? What on the food supplies? Now that much of it is sprayed from aeroplanes, we may all get the 'benefit' whether we like it or not.

One rather intriguing detail: the drinking of paraquat has now taken the place, in rural circles, of the old noose hung from a rafter in the barn for ending the lives of those who find the boredom of it all too much to put up with any longer.

103

People, Plants and Soil
―――― Herbert Girardet ――――

The Bayrische Wald is a large stretch of forest in West Germany which straddles the Czech border. It contains one of the few remaining areas of primeval forest still in existence anywhere in Europe. The remoteness of the area and its hilly terrain made it unsuitable for agricultural development. Today part of it has been designated a national park and this is visited every year by many thousands of people. They are made aware by forest guides that once much of Europe was as densely forested as the Bayrische Wald.

A walk through the ancient forest in the autumn is a breathtaking experience. Huge beech trees and silver firs stretch their branches towards the clear blue sky, the shade of the forest is permeated with flickering sunlight. This forest does not have the military look of a modern conifer plantation where all the trees stand straight and to attention in their green camouflage, lined up in neat rows. The Bayrische Wald contains trees of all ages and sizes; quite often one can find that a fir tree has put down its roots right next to a mountain ash, the two almost intertwined. There is nothing regular about the configuration of trees in this forest yet there is no feeling of chaos.

We saw the yellow beech leaves being blown off their twigs by gusts of wind and slowly floating down to the forest floor to join previous generations of leaves. We looked at the thick layer of slowly decaying leaves deposited over many years which covers the forest floor like a blanket. And there on the ground are the giant trunks of trees that came crashing down thirty or forty years ago. They are disintegrating slowly, bored into by woodpeckers, eaten by insects and fungi. Once they were covered with bark, now velvety moss has taken its place. The dead wood of the tree trunk acts as a seed-bed for tiny young trees that can remain in suspended animation for decades, waiting for an opening in the canopy to give them light to grow.

The forest will renew itself as long as summers are warm enough to allow the trees to grow and the rains are sweet and regular. The fertility of the soil is replenished annually by autumn leaf-fall. The tree roots permeate the humus layer created by the leaves. The forest is a self-perpetuating, self-fertilising ecosystem which both supports and depends on the interaction of a great variety of plants, animals, fungi and bacteria.

'For soil to remain fertile that which has been taken from it has to be replaced completely.' This statement, written in the 1840s by the German chemist Justus von Liebig, revolutionised scientific understanding of soil fertility. The forest, whose leaf litter stays where it is deposited, conforms totally to this requirement. The nutrients taken out of the soil by the growth of the trees are put back into it, year after year. Farmland that results from forest clearance benefits from this store of fertility. But when farmland is planted with annual crops which are then removed by harvesting, the resulting deficit of nutrients impoverishes the soil.

Self-sufficient smallholders can maintain soil fertility rather as trees do. The crops they grow are consumed on the farm, and the waste is then returned to the fields. But commercial farmers who sell their crops cannot maintain the fertility of their land without importing plant nutrients from outside. If they do not do this their fields will stop yielding before long. This was true during the days of the Roman latifundia and it is true now.

Man 'in paradise' lived as part of a self-perpetuating ecosystem. But farming imposed on us the need, indeed the duty, to regenerate the fertility of the soil in our charge. Farmers have wrestled with this problem from the earliest days of agriculture. The Greeks and the Romans let the fields lie fallow in alternate years. The Franks and their successors introduced three-field rotations. Later, four-field rotations were practised in order to regenerate soil fertility more effectively. But no man-made ecosystem is as successful in this as the forest, a self-contained, self-fertilising system.

Liebig conducted his research into plant nutrition at his laboratory at Giessen, near Frankfurt, which has been preserved as a museum. A protégé of the German explorer and botanist Wilhelm von Humboldt, Liebig became a professor of chemistry at the age of twenty-one. For many years he was largely concerned with analytical chemistry but the famines which reoccurred in the 1840s focused his thinking on the practical question of plant nutrition. At his laboratory, which today seems like a medieval alchemist's kitchen, he incinerated plant materials in pottery containers in order to be able to analyse the ash and the gasses which resulted. He showed that they are made up of about fifteen chemical elements which they utilise in the growth processes.

The elements Liebig found in the plant ash were nitrogen, phosphorus, potassium, sulphur, calcium, magnesium, iron, manganese, boron, copper, zinc and molybdenum. (Others have since been added to this list.) Liebig reasoned that these elements must have originated from rocks, minerals and organic matter that contributed to the formation of the soil. When crops are harvested and removed from the field on which they grew, the elements which the crop secured from the soil are likewise removed and the field is depleted to that extent. Most elements are removed in very small quantities. Only in the case of phosphorus, potassium, calcium, magnesium and nitrogen do shortages develop quite rapidly in most farm soils. It was Liebig who discovered that these elements had to be replaced annually in order to secure satisfactory plant growth. With this discovery he laid the foundation of contemporary farming practices.

The rapid growth of cities caused Liebig to become greatly concerned for Europe's soil fertility. London expanded faster than any of them.

Population growth, the enclosure of farmland and the industrial revolution had brought about this expansion of cities. Liebig wondered what would happen to Europe's soil fertility if the plant nutrients exported to the cities in the form of foodstuffs were not returned to the land. He came to the conclusion that Europeans were engaged in a process of ruinous exploitation of the soil, dating back to Greek and Roman days.

Liebig pointed to Chinese and Japanese agriculture as examples we should follow. Farming in both these countries was based on waste recycling through which soil fertility could be sustained. The nutrients that were taken out were put back. The Chinese and the Japanese returned human excrement to the soil, the Europeans flushed it down the sewers into the rivers and the sea.

In the 1840s Liebig paid two visits to London to persuade the authorities against the wasteful dumping of sewage in the North Sea. He gave public lectures and he wrote articles in scientific journals against this plunder of soil fertility. He stimulated a debate about better use of sewage but in the end it was to no avail. The huge deposits of guano on islands off the coast of Peru were discovered at about this time and soon these were shipped in great quantities, first to Britain, and then also to Germany and the USA. It is ironic that the droppings of South American seabirds, feeding on anchovy shoals far away in the Pacific Ocean, came to the rescue of Europe's soils.

Liebig came to realise that urban sewage disposal was here to stay and that Europeans would not adopt the habits of the Chinese. After all, excrement is an embarrassing topic to people in much of Europe. Liebig also realised that the guano deposits would not last for more than a couple of decades. He set to work in his laboratory to develop mineral fertilisers which could be applied to the soil in granular form. Their main constituents were phosphate and potash. Liebig was aware that plants need considerable quantities of nitrogen for their growth, but he was convinced that they could absorb most of this from the rain and from the soil itself, which usually had large stores of it.

The fertilisers didn't work and Liebig could not understand why. It took him decades to find out the reason: they were not water-soluble and so could not be absorbed by the pumping action of the plant roots. When he eventually made his fertilisers water-soluble they succeeded in making the crops grow. Thus the age of mineral fertilisers had dawned, the contemporary agricultural revolution was set in motion.

John Lawes at Rothamstead in Britain, a pragmatic experimenter, was the most successful pioneer in the actual use of mineral fertilisers. In 1842 he patented a production process for super-phosphate. He also used ammonia (a nitrogen compound), and potash and proved on his trial plots that they could considerably increase grain yields. The continuous field-trials of wheat, barley, roots and grass, which he initiated, continue at Rothamstead – now a government-funded research station – to this day. They show that it is possible under certain circumstances to grow the same crop on the same field year after year by replacing the nutrients taken out of the soil in accurate quantities. But the most successful plots are those in which farmyard manure is applied, by itself or in combination with mineral fertilisers.

In his later writings Liebig stressed that it had taken him a long time to understand that the soil particles themselves attracted plant nutrients, 'like a magnet attracts iron filings and holds on to them so that none of the particles can get lost'. He expressed the conviction that the soil is like a large filter; it ensures that soil water containing decaying organic matter is cleaned totally by trickling through it. Mineral fertilisers would be absorbed by the soil in a similar way and would thus become available to the plants as nutrients. He stressed repeatedly that only a full understanding of the reality and the laws of nature could be a sufficient basis on which to develop and improve agriculture. There is little doubt that farming today would not be the same without Liebig's pioneering research. He is claimed as a founder father by both the 'chemical' and the 'biological' schools of farming.

Nobody disputes now that the plant nutrients that have been removed from the soil by plant growth have to be replaced in full but there is sharp disagreement about how this should be done. Throughout the industrialised world, mineral fertilisers have taken over as the main source of plant nutrients and there is no doubt that – with the additional factor of improved crop varieties – average yields have gone up considerably. But at what cost? How effective is the alternative, 'biological', approach?

At the end of the nineteenth century interest in all the manifestations of the living world grew rapidly. It was the distinguished scientist Ernst Haeckel who coined the term 'ecology' in 1873. And it was Charles Darwin who conducted the first comprehensive study of the life-cycle of earthworms, 'the tillers of the soil'. In his classical work on worms, published in 1881, he said: 'It may be doubted whether there are many other animals which have played so important a part in the history of the world as have these lowly organised creatures.'

Worms burrow through the soil and provide aeration and drainage channels; they increase the organic content of the surface layers of the soil, thus favouring plant growth; they increase the depth of the topsoil by burrowing into subsoil; wormcasts are an ideal medium for the germination of seeds; worm-burrows are lined with mucus and worm-casts which are utilised as nutrients by the roots of plants.

It was the biologist and philosopher Raoul Heinrich Francé who first revealed the wondrous world of life in the soil. Born in Hungary in 1874 of a French father and a Bohemian mother, Francé developed an early interest in the life sciences. By the time he was twenty-one he had already conducted major studies of lake ecology and moorland ecosystems. In 1906 he founded the *Deutsche Mikrologische Gesellschaft* in Munich and set up a research institute. With the help of the microscope (by then well-developed), he set out to investigate life in the soil. What he found quickly intrigued a great many people, for he published his findings in books and in popular magazines such as *Mikrokosmos*. He took photographs through the microscope of a great many minute bacteria, fungi and insects which live just under our feet but had never been seen before. The gardener who digs up the garden with his spade will soon come across wriggling earthworms, but most of their fellow creatures in the soil remain unseen. Worms, algae, insects, fungi, bacteria, rhizopods – they all live down there in their underground world and without them plants and animals on the surface would not be alive. They bind the minerals that feed the plants that feed us.

Francé found hundreds of nameless little creatures, some of them so small that even the best optic microscopes could hardly reveal them. For years he analysed soil samples from many places to obtain an understanding of the great variety of life that exists in the soil. Most of the creatures he found were without vision since they exist in permanent darkness. Many died when exposed to severe frost but they left eggs or spores behind which came to life when warmth and moisture came back.

Francé gave the many varied forms of life in the soil the collective name *edaphon*.

Here seems to me the point where the existence of the edaphon is particularly important for the farmer. The soil organisms play a crucial role in breaking down organic substances and aerating the soil; they are indispensable for the metabolism of a vital soil. Whatever methods we find by which their quantity is increased will be important for agriculture for they will stimulate soil fertility. The edaphon seems to be no less important for humus formation . . . The humus substances generate carbonic acid and cause undissolved minerals to be solved or weathered.

This is the starting point for biological or organic agriculture which is now being studied with rapidly-growing interest by farmers, scientists and or more earthworms and uncounted billions of other soil organisms. It is considered vitally important that these are given the best possible living conditions because they will stimulate soil fertility and the growth of healthy and bountiful crops.

Francé and his successors acknowledge the importance of Liebig's discovery that plant nutrients removed by plant growth must be replaced in full. They also emphasise his statement that soluble minerals stick to soil particles as if to magnets. But they argue that there are better ways of supplying nutrients to plants than using mineral fertilisers. Recycled organic waste such as compost, farmyard manure and neutralised sewage slurry are considered the best substances for maintaining soil fertility: only if they are deficient in minerals such as phosphate then these might be added in the form of mineral fertilisers.

Hardy Vogtmann, the first professor for alternative agriculture to teach at a European university, is convinced that everything possible must be done to stimulate soil life and to increase the humus content if we want our soils to yield long into the future. After having been director of the Institute for Biological Agriculture in Oberwil, Switzerland, for some years, he now teaches at the agricultural faculty of Kassel University in West Germany. In July 1984, at a newly-established research farm near Witzenhausen right on the East German border, he told us what he has in mind.

Soil is not simply a substance that plant roots should be able to hold on to and to which a quantity of fertilisers should be added so that the plants can grow. It is essential that we get a better understanding of the soil. We have to realise that soil is a very sensitive system in which the soil life plays a crucial role. In our present approach to farming we have largely replaced biology with technology. We must develop biological systems of farming with which we can achieve high yields whilst reducing environmental impact. Technology should complement biology, it shouldn't replace it.

Vogtmann and his colleagues and students at Witzenhausen have set up field-trials in which they study effective ways to stimulate the micro-organisms in the soil and to increase its organic content. Leguminous crops like clover and alfalfa are much in evidence. These supply nitrogen to the soil and also provide nutritious feed for the animals on the farm. When grains like wheat or barley have been harvested the soil is sown with a cover crop, such as alfalfa or mustard, which can later be ploughed in as 'green manure'. Every effort is made to recycle all the waste which accumulates on the farm. Animal manure is composted with straw and regularly tested for its content of plant nutrients. The large compost heaps along the edges of the fields of the research farm are loaded on to specially adapted muckspreaders using fork-lift equipment. The soil is tested regularly to assess the effect of the application of the compost.

Vogtmann and his colleagues have initiated an experiment to recycle the

Vogtmann and his colleagues have initiated an experiment to recycle the household waste of the local town of Witzenhausen. Householders have been given two dustbins, a grey one and a green one. The green one is used to store all organic wastes – paper, cardboard, kitchen scraps, hedge clippings, etc. – which are collected once a week. This material is deposited on a site near the research farm and breaks down into nutrient-rich compost suitable for use on farms and market gardens. Detailed studies of this system of waste recycling have shown that it can actually make money for urban authorities, dramatically reducing the overall costs of waste disposal while providing good-quality compost for local farmers.

New methods of pest control are also being studied at Witzenhausen. These are based on the initial assumption that crops growing on a healthy soil, containing an active soil life and all the trace elements necessary for the crops, should give plants a better resistance against disease attack. Most insect pests that attack the plants are held in check by natural pest predators which are given suitable habitats. Some insect pests are kept under control by newly-developed traps to which they are attracted by the scent they give off. Some non-persistent sprays based on plant substances are used occasionally to kill harmful insects.

Weeds in the fields are kept down by newly-developed, tractor-drawn mechanical hoes and by using tractor-mounted flame-throwers. Both these methods have already proved to be effective and economically viable. In addition, crop rotations are used to ensure that constant changes in the living conditions of the weeds prevent them from becoming persistent pests.

The Witzenhausen experiments are being watched with great interest by a growing number of farmers and agricultural researchers all over Germany. Consumers are becoming increasingly concerned about possible pesticide and nitrite residues in the food they buy and there has been a massive increase in demand for biologically produced foods. Well over 1500 farmers in Germany have now gone over to biological production methods and their numbers are increasing.

The biological agriculture movement is becoming highly effective all over Europe. This is not only because people are concerned about what farm chemicals may end up in the food on their dining table; more and more people are also becoming worried about what 'conventional' farming methods are doing to the soil, streams, rivers and lakes.

We went to see the chemically-controlled vineyards on the banks of the Rhine in the area around Nierstein. The vines are usually planted on south-facing slopes to obtain the full benefit of the light and warmth of the sun. On a hot day in July, the dark-green foliage was already dotted with bunches of grapes. Tractors were going up and down the rows, spraying mists of fungicides and herbicides. The soil was bare, not a weed in sight.

A few miles away down in a valley we saw the construction of a pipeline. The site was busy with huge diggers cutting a trench into the field, lorries delivering and unloading pipes, workmen smoothing the surface of the trench and shovelling in sand. The cast-iron pipes were being lowered, put into place and linked together with plastic sleeves.

This pipeline was being constructed to supply Nierstein and other local villages and towns with drinking water. The well water in this area has such a high nitrate content that it is considered dangerous to the health of babies, young children and pregnant women. The nitrate comes from the fertiliser applied to the vineyards and grain fields. The lifeless soil and the vines cannot absorb all the mineral fertiliser applied to it and as much as half the nitrate ends up in the groundwater. One well near Nierstein was found by water authority officials to contain eight times the amount of nitrate recommended as safe by the World Health Organisation.

There is no doubt that many more such pipelines will have to be laid all over Europe, if the supply of safe drinking water is to be assured in areas where intensive agriculture is practised. Nitrate accumulation in groundwater, in streams, rivers and lakes is becoming a widespread problem wherever vines or grains are grown on a large scale. The wheat-growing areas on the lower Rhine are as badly affected as those in Norfolk, Cambridgeshire and Staffordshire. The problem is likely to become worse in the coming years as nitrate can take decades to find its way from the soil surface down into the groundwater. The annual nitrate application in these areas is now over 89 lbs per acre, 100 kilograms per hectare, and even though half of this may not be utilised by the plants, farmers are still prepared to continue at this rate to produce maximum yields. The removal of nitrate from water is a very costly business: with present technology it costs around £500 per hectare per year to make the water safe which is, of course, totally uneconomic.

Professor Mengel, whom we interviewed in July 1984 at his laboratory at the Justus von Liebig institute for plant nutrition at Giessen University told us, 'There is often a tendency to apply more nitrate then necessary because its impact is very apparent in plant growth . . . We do not deny that considerable nitrate leaching can occur. But we also say that we can counter this problem if we examine our soils more carefully for available nitrogen.' However, he does not believe that farmers today can do without mineral fertilisers. He sees it as extremely difficult to return all the fertility removed from the farm to urban markets back to the fields. After all, 90 per cent of food consumers in Europe live in towns and cities. Household sewage is mixed with industrial sewage so that potential plant nutrients are contaminated with chemicals and heavy metals. Professor Mengel expressed concern about the waste of phosphate in our sewage systems.

He stressed that phosphate is likely to be in short supply in two or three hundred years. There will be no replacement for the deposits now being mined. He emphasised that we must therefore reclaim the phosphate which at present goes into the sewage treatment plants. There is no doubt that if society had the will to tackle the problem of nutrient loss, appropriate methods could be found to make this a viable proposition.

Professor Mengel and many of his like-minded colleagues in Europe and America now accept that mineral fertilisers have worrying environmental impacts, but they would argue that since we cannot do without them, we should make better use of them. For example, he says that cover crops and green manuring help the land to recycle nutrients and to absorb mineral fertilisers more effectively.

Those who are against feeding farm crops from the fertiliser bag, and Professor Vogtmann is one of these, would argue that mineral fertilisers have *many* side-effects in addition to the problems of water contamination. They argue that soil regularly dosed with nitrate becomes dependent on receiving a constant supply. This has been confirmed in recent research. Microbes in the soil that assimilate nitrogen from the air are suppressed by mineral nitrate. They are greatly diminished in number and therefore make crops dependent on ever greater applications of bag fertiliser.

Vogtmann and others have shown that green vegetables, particularly if grown in greenhouses during the winter months, can accumulate excessive amounts of nitrate in their tissue which, if converted into nitrite, can become a health hazard. They stress that the greater weight of vegetables grown with mineral fertilisers is often caused by an excessive accumulation of water in the plant tissues. This does not contribute to the food value of these vegetables. On the contrary, it reduces their storage life compared with compost-grown vegetables. Some supermarket chains have now accepted this argument and have begun buying and selling organically-grown produce.

Mineral fertilisers, in association with pesticides and soil compaction caused by heavy tractors, can contribute to soil erosion. A reduction in the variety and quantity of soil life, which can be caused by the combined effects of 'industrial' farming, has now been found to occur in many different locations. We have seen that a healthy, 'living' soil has a loose, crumbly structure which can absorb water like a sponge and is able to retain it for later use by the crops. Such soil is friable and easy to plough. When it becomes dry during prolonged periods of low rainfall or drought, it still holds together and does not blow away easily.

'Lifeless' soil, on the other hand, compacts easily. It requires stronger and larger tractors to plough it. Heavy rain usually reduces the soil surface to an impervious pan. The rain finds it difficult to penetrate this and as a

112

Above, *Irrigation wheels on a farm in Idaho, America. Huge acreages can be irrigated by using groundwater supplies – but for how long?;* below, *Contour ploughing in Idaho helps to arrest erosion;* overleaf, *The polders of Holland where detritus washed down the Rhine from the Alps has been turned into farmland*

Organic farming in Germany: Above, *Hand-weeding;* below, *Turning household waste into compost by using a muckspreader*

Previous page, *Farming in China: new terraces have been built to prevent soil erosion and to increase the area under crops;* inset, *Vegetable fields in Canton, where intensive cultivation produces some of the highest yields achieved anywhere*

Neukirchen am Grossvenediger in the first Alpine National Park, Austria. The government authorities and local farmers and villagers have formulated an ecologically-benign development plan, including low-input farming, traditional crafts and waste-recycling technology, all of which reduce the environmental impact of farming and tourism

result runs off on the surface or forms puddles. On slopes, bad soil structure can quickly lead to gully or sheet erosion. This is now a recognised problem in many parts of Europe, particularly in areas where crops such as maize are grown which leave the soil exposed for long stretches of time.

Opponents of farming methods which totally depend on mineral fertilisers also emphasise that crops grown in this manner are usually less drought resistant. Their roots are shallow since the fertilisers are available to the plants just below the soil surface. More irrigation is necessary which in the hotter countries can lead to salinisation – the build-up of salt in the soil.

Finally, it is argued that nitrate-based fertilisers are very costly in energy: around two tons of oil are required to produce one ton of nitrate fertiliser. As energy prices go up, nitrate prices must inevitably follow suit. How much longer will farmers be able to afford the price?

Nitrate fertiliser began to flood Europe and America after the Second World War. They were made in the same factories that had produced nitrate-based explosives used in ammunition. They were one of the major cornerstones in the industrialisation of agriculture which occurred in the 1950s and 1960s.

In the European Community, the mechanisation and chemicalisation of farming were pursued with vigour by politicians and industrialists. For the first time, farmers became important customers for industrial products. Labour in rural areas which became surplus to requirement was absorbed by the rapidly-growing industries all over Europe.

As it became possible to achieve higher yields, the use of manure and the age-old practice of mixed farming was abandoned in many places. Monocultures of grain, maize, sugar-beet, potatoes and other major farm crops were introduced in most areas where large-scale arable cultivation is possible. Thus, use of ever more powerful tractors, of ever larger and more sophisticated farm machinery, raised labour productivity on the farms to previously unknown levels, displacing millions of farm workers and small farmers.

At the beginning of the nineteenth century in Germany, four people were fed by one farmer. Today a farmer feeds forty-two people. In Britain the figure is about eighty.

In the EEC, the commissioner for agriculture, Sicco Mansholt, formulated a plan which became the basis of the agricultural policy. This was his rationale, set out in his pamphlet, *The Common Agricultural Policy: Some New Thinking*, published in 1979:

Opposite, Decorated cows coming down in the autumn from the mountain pastures in Austria. This is a festive occasion in mountain villages throughout the Alps

121

If we are to provide the same conditions for those working in agriculture as the rest of society in respect of leisure time, holidays, income, etc., it is necessary substantially to increase the *per capita* productivity. This implies the use of artificial fertilisers and pest control preparations and an increased size of holding. It is evident that on a holding of some twenty cows and a few pigs it would not be possible to achieve those social conditions for the farmer that, say, a factory worker enjoys.

The subsidies paid to farmers in the Common Market under the CAP have been a subject of lively debate for many years. Their main purpose was to prevent price fluctuation and provide an assured income for farmers. There is no doubt that the largest farmers have benefited the most, since subsidies are paid on the volume of produce delivered. The big farmers were put in to a position to squeeze out smaller neighbours. Farm amalgamation was the order of the day.

In the 1960s and 1970s the rationalisation of landscapes throughout Europe had a dramatic effect. The drainage of wetlands, the removal of hedgerows and woodlands became causes for public concern. Environmental groups were formed to campaign against the 'vandalisation of the landscape' with the help of public funds. In Britain alone, 140,000 miles of hedgerows were grubbed up in forty years, vastly reducing wildlife habitats in agricultural areas. In recent years, protests have become increasingly passionate. People are becoming convinced that the environmental and social costs of a 'modern, efficient' agriculture are too high.

Sicco Mansholt himself, now retired, has expressed grave concern about the effect of the policies he initiated. There is no doubt that it will take great effort to bring about a major change of direction. But the need for change is now there for all to see.

By the mid-1970s industry could no longer absorb the surplus labour which had been made redundant from the farms. When unemployment became endemic in both town and country, many people began to see the plans of the EEC to make farming more labour-efficient as counterproductive. Yet urban food consumers and governments still begrudge small farmers the public money that would allow them to stay in business.

In the last few years it has become very apparent that industrialised farming is environmentally unacceptable. We have already seen how this applies to the effect of mineral fertilisers on groundwater, streams and rivers. The separation of arable farming from animal husbandry is another case in point. Concentrated animal rearing in huge sheds is now commonplace throughout Europe. The accumulation of slurry in these units is becoming a widespread problem. The region around the town of Vechta in Lower Saxony, West Germany, is a good example. It is 'famous' for its concentration of 'animal-rearing factories'. Much of the feed fed to the animals is brought in from the USA. In their book *Die Lage der Nation*, Edmont Koch and Fritz Vahrenholt describe the problem:

The Vechta district has a population of 15 million chickens, over 83,000 beef cattle and more than 600,000 pigs, in addition to whole armies of ducks, turkeys and geese and, last but not least, 97,000 human beings. The chicken produce 2000 tons of droppings a year, and the cattle and pigs produce a similar quantity of slurry ... The Vechta district should have a 'manure disposal area' of much more than 80,000 hectares, following official recommendations. But the total area of farmland in the district is only about 57,000 hectares.

It is thought that only about 15 per cent of the nitrate deposited on the fields in the form of liquid slurry is absorbed by the plants during the winter months, 65 per cent ends up in the atmosphere and – even worse – 20 per cent finds its way into the groundwater. As a consequence, nitrate levels in the water exceed the permitted limit in two-thirds of the 300 samples taken at the end of 1982. When new, safer limits are introduced shortly, it will be very difficult indeed to find clean water supplies which can be blended with the contaminated drinking water.

In the Vechta area the owners of some of the largest animal units have rented fields from farmers for the exclusive purpose of slurry disposal. Here crops are no longer grown at all, not even maize which is best able to cope with large doses of slurry.

The case of Vechta is a dramatic example of a trend which can be seen all over Europe and America: too many animals on too little land. Contamination of groundwater supplies by nitrates from slurry is now a widespread problem. It is not only drinking water which is affected but also water in rivers and lakes. The combination of nitrate and phosphate is particularly damaging. Slurry, sewage and mineral fertiliser run-off from fields has caused the virtual death of many hundreds of lakes all over Europe. Very expensive remedies have to be found to undo the damage.

An interesting example of a dying lake is the Baldegger See in Switzerland, which is surrounded by small farms specialising in fattening pigs and cattle. A considerable proportion of the feed used on these farms is bought in from the USA and elsewhere. Here, too, the animals produce more slurry than the land can cope with. Nitrate and phosphate seep down into the groundwater or run off into streams and find their way into the lake. For years it was eutrophied: the phospate contained in the water stimulated the excessive growth of algae, much more than the lake fauna could feed on. The decay of the surplus algae absorbed large amounts of oxygen from the water, leaving it highly oxygen-deficient. This in turn caused the fish population to diminish dramatically. Today, the lake floor is covered with a thick layer of stinking mud which will no longer support life.

A few years ago it was bought by a Swiss conservation organisation which has set itself the task of reviving the beautiful but sick lake. Technicians have now installed a system of pipes connected up to an oxygen tank on the shore. £63,000 a year is spent on pumping oxygen into the lake in order to bring it back to life. The fish population is slowly increasing again but, as a local fisherman said, 'We catch *very* expensive fish in the Baldeg-ger See.' Hundreds of lakes throughout Europe require similar treatment but who is going to foot the bill?

For many years the Rhine has had the reputation of being the most polluted river in Europe, perhaps in the world. Gushing down from the glaciers and mountain forests of the Swiss Alps, it provides drinking water for millions of people who live in the cities, towns and villages along its banks, all the way from Lake Konstanz to the North Sea. The same people who use the Rhine for their drinking water also use it as a sewer for the disposal of liquid wastes.

As recently as a hundred years ago, the Rhine was immensely rich in fish, providing large quantities of protein for the people who lived on both sides of the river. Fifty different kinds of fish used to inhabit the Rhine whose grassy banks and quiet sidewaters were ideal spawning grounds. In 1875, in one Dutch fish market on the Rhine, 56,000 salmon were sold, which is thought to have been just over half the total sold in Holland that year.

Since 1880, reductions in fish quantities were observed, particularly salmon and sea trout which would swim up the river to spawn down-stream from the Rhine falls at Schaffhausen close to the Swiss-German border. But even around the turn of the century, servants in wealthy households along the banks of the Lower Rhine insisted it was written into their contracts of employment that they should not have to eat salmon more than three times a week.

Today the banks of the Rhine are the location for a vast assortment of factories. In fact, 20 per cent of the world's total chemical production capacity is located along the Rhine, ranging from drug factories in Basle to refineries in Alsace, the giant Bayer works in Leverkusen and chemical plants in Holland.

Chemical pollution of the Rhine from industrial sources probably reached its climax in the mid-1970s when widespread anxiety about the condition of its water led to a public enquiry. Since then the International Commission for the Protection of the Rhine against Pollution has pub-lished an annual report, sponsored by the Swiss, German, French and Dutch governments. This report lists the total content of the river from industrial, mining, household and agricultural sources. These include organic and inorganic chemicals, metals, and eutrophying substances like phosphates and nitrates. A chain of warning stations along the banks has been set up which is meant to forewarn people of oil and chemical spills from riverboats and factories on the Rhine. Increased monitoring has led to a reduction of irregular chemical pollution of the river.

In the 1950s and 1960s, reports of dead fish in the Rhine were regular items in the German and Dutch press. Living conditions for fish were made increasingly difficult because of pollution, removal of vegetation along the riverbanks and ever heavier shipping traffic. From 1960 onwards, commercial fishing in the river ceased altogether. The loss of fish production in the German section of the river which was valued at £2. million in 1955 was considered insignificant compared with the benefits from increases in industrial production.

From the mid-1960s, fishermen were no longer allowed to sell fish caught in the Rhine, and in Germany this is still the case. High heavy metal contents and accumulation of dangerous chemicals (HCB, PCB) in the fatty tissues of the fish were the reasons. In recent years, treatment plants built for settlements and factories have reduced the load of chemical carried by the Rhine and the oxygen content has gone up a little. But the Rhine is still a grossly polluted river. The Rhine carries with it not only the pollutants from factories but also the wasted plant nutrients from a huge area of agricultural land, and effluence from many cities, towns and villages.

When looking at the statistics one's attention is drawn to the dramatic increase of *nitrate* in the Rhine which has occurred since measurements were first taken in 1972. In 1981 at the measuring point on the German-Dutch border the Rhine carried a total nitrogen load of 355,000 tons. This would be sufficient to supply the whole of Dutch agriculture with nitrate fertiliser if it were possible to divert the Rhine across the fields of Holland! The total load of phosphorus that year was 21,000 tons. The effect of this massive load of plant nutrients on the North Sea is still not fully understood but there is evidence that over-production of plankton does occur in parts of the North Sea, notably the German Byte which also receives nutrient loads from the Elbe and the Weser. The thick layers of froth which have been observed on the beaches of North Sea islands such as Helgoland in the summer months are thought to be due to excessive growth of plankton caused by the nutrient loads of these rivers.

It is clear that western society is causing the waste of valuable plant nutrients on a gigantic scale. We refuse to recycle our household wastes in the manner advocated by Liebig and allow them to become pollutants. We apply excessive amounts of mineral fertilisers to our fields, causing long-term pollution to groundwater, streams and rivers. We cause the deterioration of soil structure by the same process, undermining the inherent fertility of the land. We separate animal husbandry from the land which has resulted in accumulation of excessive amounts of slurry which have become pollutants of soil and water.

What is the sense of it all? How better can we undermine the future of human existence?

A Look at North America
———— John Seymour ————

I sit next to the driver in a large American car somewhere near the city of
Davis, in northern California. The driver is showing me his tomato farm.
We drive, quite fast, for what seems to me to be mile after mile of dead
straight road, and at right angles to the road run equally dead straight rows
of rather stunted-looking tomato plants, as far as the horizon on both sides
of us. The plain over which we are travelling is dead flat. Except for the
road ahead, the rows of tomato plants, and the sky, there is absolutely
nothing to be seen. I note there are no birds. Also there are no weeds. There
are no trees. Is this a vision, I wonder, of the future?

Occasionally there is a blank where a row of tomatoes should be. 'Is that
where the seed drill bunged up?' I ask. 'No,' comes the reply. 'It's where the
fertiliser spout bunged up.'

Nothing, he told me, would grow on that land without fertiliser. The
land, which originally had been wonderfully fertile, was inert.

'Do you put nothing back into it?' I asked. 'What happens to the
haulms?'

'Burn 'em. Disease.'

'What about steer manure?'

Now steer manure is a substance in almost unlimited supply in the
western United States. For most of the cattle of North America end up
being fattened in beef lots. Most of these are enormous, often containing
thousands of cattle. Needless to say, the manure from these animals builds
up into mountains. Anybody can go and help themselves to it, for free, as
much as they can carry away. Yet all but a tiny proportion of it just rots
and rots away in those huge heaps and eventually disappears. It must be
the most grandiose waste of a valuable resource that has ever occurred on
this planet.

'And how could I pay the labour to fetch and spread the stuff?' he
replied.

Of course the worst crime of so-called modern agricultural economics
has been to divorce the farm animals from the land. I cannot help quoting
from a magnificent book written by a Kentucky poet and farmer, Wendell
Berry, called *The Unsettling of America,* in this context.

Berry quotes from an article written by a man named Jules B.Billard in
the February 1970 edition of the *National Geographic.* I read the article
when it came out and was amused by this fine example of 'gee-whizz

journalism'. It is about the future of agriculture in the United States. Mr Billard admits that the confining of all cattle on specialised 'beef-lots' is going to cause an enormous environmental problem. (Indeed one already exists – the problem of what to do with the mountains of raw manure and lagoons of stinking slurry is already insoluble.) He also notes (with approval) that in 1968 American farmers spread 'nearly 40 million tons' of chemical fertilisers on their land (an enormous expense and another dangerous source of pollution). Wendell Berry notes that the connection between these two severe problems is completely missed, 'Mr Billard forgot, or he never knew, that once plants and animals were raised together on the same farms – which therefore neither produced unmanageable supluses of manure, to be wasted and to pollute the water supply, nor depended on such quantities of commercial fertiliser. The genius of American farm experts is very well demonstrated here: they can take a solution and divide it very neatly into two problems.'

We drove back to the tomato-grower's farmstead, past the wooden sheds where the 'wetbacks' lived (Mexican labourers are called wetbacks because they have to swim the Rio Grande to enter, illegally, into the United States), and we went into his enormous machinery shed. It was nearly filled by two colossal machines. One of these was brand new, the other three years old. Tomato harvesters.

'I paid 35,000 dollars for that one three years ago,' he said. 'Now it's obsolete. Had to buy this one – it set me back 60,000 bucks. The old one needed twenty wetbacks riding on it to sort the tomatoes. The new one has an electronic eye that can grade according to colour. It only needs five people to operate.'

'What will you do with the old one?'

'Bust it up. Nobody wants it.'

He went on to tell me that there was no way, as long as he could live, that he could ever get out of debt. His debts would always exceed the value of his farm, so even if he sold everything he would still be in debt. The only thing he could do was to struggle on, in debt, being carried by the packing firm which bought his tomatoes for the rest of his life.

On the way from the machinery shed to the house we passed through a fine vegetable garden, and in it were growing some most healthy and excellent-looking tomato plants.

'Why?' I asked, 'Why? Why do you bother to grow tomatoes in your kitchen garden when you have – how many thousand acres? – of the dam' things growing just out there?'

'Do you think I'd eat those things?' he said. 'If you saw the poisons we put on them you wouldn't ask that question. You'd never eat another canned tomato in your life.'

And he gave me a run-down of the poisons he put on the tomatoes outside. Some of them were systemic poisons, that is, the poisons which enter the plant so that every cell contains some.

'And why don't you have to put all this poison on these plants?' I asked.

'Just what you were talking about,' he said. 'Steer manure. Steer manure. These plants have got resistance.' And he went on to say that his garden tomatoes were grown on fresh ground every year. The field tomatoes outside were grown on ground that had been growing the same crop for twenty years.

I asked him if he ever thought of growing a break crop outside.

'Can't afford to. The only way I can meet my interest and mortage payments and pay the labour is to get a full crop of tomatoes every year. And we can do it – we can do it with chemicals. More and more chemicals.'

Yet the chemicals were getting more and more expensive, he added. Looking over his fence at the vast plain covered with stunted plants I thought to myself that, whatever this worried man might be farming like this for, it certainly was not for *pleasure*.

In the autumn of 1984 we drove to an area called the Wolf River District. I wished to see this, the heart of the mid-west of America, because it is the area which is at present absolutely crucial to mankind.

The only country in the world that has a larger area suitable for growing wheat than the Great Plains area of North America is the USSR. The 320 million acres, or 130 million hectares, of black earth country in the Soviet Union now produces four-fifths of the country's food but this soil is deteriorating rapidly. According to the *Observer* in February 1985, an article in the Russian ecological magazine *Man and Nature* makes the point that, 'If we do not take rapid steps to preserve the black earth then in fifteen to twenty years an irreversible process will begin and then there will be nothing to save.'

Besides the 320 million acres of this once-fine black earth the Soviets have 80 million acres of land that has been reclaimed (mostly by irrigation) since the Revolution but this, owing to bad farming, is rapidly eroding away.

The result is that in 1985 the Soviets have offered to buy 40 million tons of United States wheat which will comprise quite a large component of their total wheat demands. Over the past two decades Russia has become increasingly dependent on United States grain. Wheat is also pouring into Russia from the Argentine, Australia and other countries but it is quite obvious that if North American grain were cut off the Russians would begin to starve.

The Third World, too, is becoming more and more dependent on United States grain. As territory after territory in Africa and Asia reaches famine conditions, grain aid is poured in, and most of it from North America.

And even western Europe, in spite of what it now calls its wheat mountain, is far more dependent on the American Great Plains than most Europeans like to think. As every year goes by, American maize and American soya-beans form a bigger component in European livestock rations. Europe has a wheat mountain only because it is importing vast tonnages of North American beans and grain.

So what would happen to the world if North American agriculture failed? The possibility just does not bear thinking about. Russia would starve. Europe would at least go hungry. The famine areas, which increase every year, would go without any relief at all. And the North Americans would fall on very hard times. The enormity of the consequences is why I decided to go to Middle America to see for myself.

I drove out from Kansas City through miles and miles of country very much spoiled by its proximity to a great city. When I got to the country proper, I found it gently rolling, treeless, with most of the land under the plough. There were good-looking, white-painted, weather-boarded farm-houses scattered about, next to big barns and grain silos.

The land was a succession of bluffs separated by small valleys. In places, reefs of limestone rock were exposed, reminding me of the Cotswold oolite. The soil was obviously very deep and much of it was loess – wind-blown soil. After the last Ice Age, the sand and silt that had been eroded by ice and water from the great mountain ranges – the Laurentian Mountains to the north, the Appalachians to the east and the Rockies to the west – had drifted in the wind. There must have been an era of great winds and droughts for the wind-driven dust has been deposited in great thicknesses: I saw a soil profile, revealed by an erosion gulley, which showed loess soil forty feet thick and was told that occasionally it was even thicker. In places, the soil was not loess but obviously a glacial till: the sort of soil that I was accustomed to when I owned a farm in west Wales. This soil, too, had been brought from afar: shoved and rolled along by the glaciers of successive Ice Ages and dumped where it still lies.

On all the arable land that I saw (the month was December and the land was bare of crops), there were signs of sheet erosion, in many cases severe. At the lowest point of several fields there was quite a distinct delta of topsoil that had been washed down the hill and had fetched up there. Wherever there was a valley between two of the rounded bluffs that formed most of the landscape there were signs of incipient gulley erosion. This had in most cases been disguised by the farmer by the simple process of ploughing down the cleft of the valley. I realised that he had to do this – otherwise the gulley would soon have got so deep that he would have been unable to cross it with the combine harvester.

A small proportion of the fields on the steeper slopes had been terraced, or at least treated to a sort of bastard-terracing: not the true terraces of Italy, Greece and the ricelands of Asia, where each terrace surface is quite level and the soil kept up by well-built stone walls, but sloping terraces made by scraping soil *up* the hill into contour banks which had been subsequently seeded by grass. Sheet erosion was still going on here but on a reduced scale. In places, where land had been thus bastard-terraced, there were grassed channels for surplus water to run away to the stream below; in others, a more elaborate system of underground drains took away surplus water. I reflected that in a natural countryside, or a properly farmed artificial one, there should be no run off of surplus water. All rain, except perhaps in the most extreme of freak thunderstorms, should be absorbed by the vegetation and the humus-rich soil. But here, in winter, there was no vegetation, and a glance at the soil made one realise that it contained only a minimal amount of humus.

I drove through the small river port of St Joseph, crossed the Missouri (which was *café-au-lait* colour with the soils of the United States), drove through Troy and eventually arrived at the little town of Robinson.

This was like a movie set. It was easy to imagine the present President of the United States, in his younger, wilder days, riding down the wide main street firing a revolver into the air.

At the edge of the little town was a collection of grain elevators, and near these was a shop that had been converted into an office. The illusion that I had driven into a film set was heightened by the emergence, from this building, of a young man resplendent in the traditional uniform of the American cowboy. He invited me into the office and there introduced me to a young lady who he said was the manager of the Wolf River Watershed Joint District No.66, whose office this was. There were three other men there, too. The cowboy not only dressed like a cowboy but *was* a cowboy – or at least he owned a ranch. The other men were farmers, arable farmers. But they worked for the Watershed in their spare time because they were all genuinely concerned about their eroding country.

During the days that followed I learned something about how such a strangely-named organisation as the Wolf River Watershed had come about.

The land about there was taken from the Indians in 1854. In 1866 there was a land rush and the whole country was grabbed by white settlers. By 1911 a newspaper could report that 'where weeds grew, now sixty to ninety bushels of corn per acre grow'. (I must explain to European readers that when the Americans say corn they mean maize. I must equally explain to American readers that when the English say corn they mean wheat, barley, oats or rye.)

It must be remembered that when the land was taken away from the Indians it was still in a completely non-artificial condition. The area we were in was on the border between the Great Woods – the huge, mostly hard wood, forest that covered the eastern part of the continent and the Great Plains. The Great Plains in the east were long-grass prairie. Further west, in Colorado for example, the plains were covered with *short*-grass prairie. All of this grass was grazed by enormous herds of wild buffalo. On the buffalo preyed the Indians, who were still at an Old Stone Age level of development. They had never invented metals. They got along very well without them. The ecosystem was perfectly stable and would have gone on in much the same way for ever, or at least until some climatic or other cataclysm occurred. Over the millennia the grasses and other prairie plants, together with the dung of the buffalo, had formed a deep and rich topsoil, largely formed of humus. The mass of roots, living, rotting or dead, absorbed any amount of rain the heavens liked to send. All the water that fell on that prairie was absorbed by the soil for the use of the plants – there was no run off, no erosion. The streams and rivers ran clear. One of the great complaints of the Indians about the white man was that his activities sullied the rivers and destroyed the fish. But it did not end there, of course. The white man destroyed the buffalo and then destroyed the prairies themselves, ploughing them up to grow grain.

In a decade or two the new settlers had cashed in the fertility stored up by 10,000 years of prairie grasses. From then on, the nutritional requirements of any crop they grew would have to 'come out of the bag'. By 1927, just sixty-one years after the big land rush, a soil survey was conducted in Doniphan County (one of the three counties that formed our 'watershed') and this concluded that three-quarters of the soil of the county was highly susceptible to erosion because of its loessal soil; that the continual production of corn on hillsides was hastening the erosion, and so was the lack of a system of crop rotation; on many hill fields the land surface had been lowered as much as three feet and often even the subsoil was gone.

Eastern Kansas was largely spared by the dust bowl disaster that blew most of the topsoil of Oklahoma and surrounding areas into the Atlantic Ocean. The effect of the dust bowl was to cause the federal government to set up the Soil Erosion Service in 1936. This was followed by the Soil Conservation Service in 1937. The SCS had some success and, by 1965, 78 per cent of the farmers of the Wolf River District were co-operating with it to some extent. The SCS tried to persuade farmers to grass down steep slopes, terrace others, not plough up and down slopes but plough along the contours, and adopt other measures.

But the problem of erosion was not being solved by the efforts of the SCS alone (in fact it was getting steadily worse), and so in 1953 the Kansas

Watershed District Act was enacted, and this permits people living within a common drainage area to organise and plan improvements for the management of water resources in their area. The farmers of Kansas thereupon organised themselves into eighty-three watershed districts covering 11.5 million acres (4.5 million hectares). From what I saw of one of them I was convinced that these are good examples of real democracy, which comes from below and is not imposed from above, and that if any organisations are going to save the soils of the United States these might.

But it will take more than organisations – it will take a whole new (or very old) philosophy and, above all, quite different financial incentives.

We drove out in a four-wheel-drive truck to one farm, so that we would see a particularly nasty gulley that had developed from a pretty level field in just twelve months. The proud possessor of the gulley, and of the farm around it, seemed to me to be like the old whore – he knew what he was doing was wrong but could not give it up! He *knew* that his monoculture of corn was leading to his soil washing away. He also knew that he had a *lot* of soil – we could all see that by looking down into his gulley, which revealed a twenty-foot depth of fine loess soil over the now denuded bedrock. What was one gulley on his 320 acres? But he also knew that sheet erosion was gradually wearing away at his patrimony. But he *had* to go on with this maize monoculture. He owed such a lot of money to the bank. He *had* to meet his interest payments every month. Mr Alistair Cook, in one of his radio programmes, has said that 10 per cent of American farmers are about to go bankrupt. Our farmer did not want to be one of the 10 per cent. At the present rate of erosion, his soil would probably last thirty years. And – who knows – by then he might be out of debt and be able to go and live in Palm Springs. He wouldn't want a farm then, anyway. Now lest any reader should think that I am exaggerating this problem of soil loss in Middle America I must give some other people's opinions. Certainly no one I actually met in Kansas had any illusions about it – they were all desperately worried about it.

One method of measuring soil loss in America is by measuring the annual production of sediment in the rivers and streams draining a given area. Bulletin No.16 of the Kansas Water Resources Board, 1971, estimates that Plains sediment ranged from 1500 to 2800 tons per square mile (5.8 to 10.8 tons per hectare) per year, but that sediments from the bluffs (as the waves of the prairie are called) average a yield of soil of 4000 tons per square mile (15.5 tons per hectare) per year and have measured as much as 15,000 tons per square mile (58 tons per hectare) per year.

To bring it nearer home, to where I was, Dick Holland, an SCS geologist, identified a maximum bluff sediment in the Wolf River Watershed of 8000 tons per square mile (30 tons per hectare) per year.

Sediment studies are made by measuring the sediment that passes one particular point on a stream. They do not reflect gross erosion. A very interesting study was conducted in Ontario, Canada, by the EPA Water Quality Office, which was reported in the EPA Bulletin No. 13020 of 1 July 1971 in a paper called 'Agricultural Pollution of Great Lakes Basin'. This shows that on permanent grass land, soil loss and water loss are 'negligible' while on land continuously cropped with corn there is a soil loss of 16,800 pounds per acres (19,050 kilograms per hectare) per year.

The staid and stately General Accounting Office of the US put in a report to Congress in 1977 which stated, 'Estimates of soil losses from 283 farms GAO visited on a random basis in the Great Plains, Corn Belt and Pacific North West indicate that topsoil losses are threatening productivity.' Productivity! Apparently it does not matter at all that the very survival of the United States itself is threatened! In the same report it was stated that the Soil Conservation Service has decided that the 'acceptable' annual loss for deep soils is five tons per year per acre, and for shallow soils one ton. It also states that at least 84 per cent of typical farms are losing *over* five tons.

Now, as soil is lost, so soil is made. There seem to be as many estimates about how long it takes to create an inch of soil as there are estimators, but the redoubtable SCS claims that it takes, according to the climate and the kind of rock the soil is being made from, from 300 to 1000 years to create an inch of soil. It goes on to say that American farmers are losing an inch every sixteen years! If this is not 'deficit financing' I know not what is.

And President Carter who, whatever we think of his presidency, was a good farmer, stated, on 2 August 1979, that 'before 1935 the US had well over 600 million acres of actual or potential cropland. Since then approximately 100 million acres of potential cropland have been effectively destroyed.' The destruction goes on – but at an increasing rate.

The destruction is not new in America. From the moment that European settlers set foot on that unsullied continent, they began to destroy it. George Washington, himself a fine farmer who still retained some of the true farming lore of his Norfolk ancestors, wrote in a letter of 1797, 'We ruin the lands that are already cleared and either cut down more wood, if we have it, or emigrate to the western country. A half, a third or even a fourth of land we mangle, well wrought and properly dressed, would produce more than the whole under our system of management, yet such is the force of habit, that we cannot depart from it.'

Well wrought and properly dressed! George Washington was a scion of an East Anglian family and must have been brought up with the traditions of High Farming which had been pioneered in that country. By 'properly dressed' he did not mean having quantities of chemicals dumped on it. He meant spreading good farmyard manure on it, carefully made from the dung of cattle or pigs composted with straw in yards covered from the rain and then allowed to rot down in aerobic heaps. By 'well wrought' he meant farmed with proper crop rotations, turnips or other fodder crops to clean the land; folded sheep probably to improve its texture and add to its fertility; a regular grass and clover break to improve the *heart* of the land.

Well, the land of America gets 'well dressed' now – with 40 million tons a year of artificial fertiliser – part of which has the direct result of 'burning' (causing to be oxidised and destroyed) what little humus is left in the soil, together with over half a million tons of poisons. Half a million tons in a country as big as the United States may not sound very much but even a tiny drop of some of these biocides can kill a human being.

Alas, the dedicated and persistent work of the Watershed Districts and the Soil Conservation Service and USDA (US Department of Agriculture) and RCA, the 1977 Resources Act, is *not* succeeding in preventing soil erosion in the United States. In the Wolf River District, Dick Holland, the SCS geologist, estimates that even if over three-quarters of farmland comes under conservation programmes (and economic pressures being what they are this is most unlikely to be achieved), *still* the loss of soil will range from 2.96 tons per year per acre to 9.37 tons (1.2 tons to 3.8 tons per hectare). This would be an improvement on today but still far too high if the land is not all going to wash away.

There are only two ways by which the soil of Middle America can be saved. One is by grassing the whole of the Great Plains down or letting them tumble back to prairie, and that will not happen if the white Americans can help it. The other is to return – or go forward – to good farming.

I saw the farm of a farmer who has done just that. On his 320 acres (130 hectares) he kept 100 head of cattle. These grazed on permanent pasture, which he had established on all the steeper slopes on his farm, and on temporary lucerne or grass leys which covered a quarter of the arable land in the summer, and were yarded on straw in the winter. Besides the inevitable corn which he had to grow as a cash crop, he grew oats and barley. His cattle looked healthy and contented. His soil looked rich and stable and in good heart.

'Do you make much money?' I asked.

'No! But at least I'll be able to hand some soil over to my grandchildren. And I survive.'

Further west in the Great Plains, in what was the short-grass prairie, conditions are different. The rainfall is much lower there, and it is not so much water erosion that is to be feared as wind erosion. The Americans have practically given up talking about the dust bowl – maybe Steinbeck said the last word about it in *Grapes of Wrath*. Many city Americans believe that it is a thing of the past and that the scientists will see that it will never happen again. The scientists – the ones who work in the field – do not think this, and nor do the farmers. That mighty cataclysm is never far from their thoughts. The rate of erosion in America has never been higher than it is today. The humus-devoid soils could blow away in the first bad drought. I never saw the dust bowl (although, as a schoolchild, I probably saw one of the results of it – for the sunsets in Europe were the more magnificent for the vast quantities of dust in the atmosphere it kicked up) and so I shall quote from one of the finest and most prophetic books to be written in our century: Edward Hyams' *Soil and Civilisation*.

Less than thirty-five years after the settlement of Oklahoma, on a day of high wind from the west, a strange dark cloud hung over the city of New York and all the coast north and south of it. The phenomenon was to be repeated, but on this first occasion of its occurrence its novelty helped to impress the ten or twelve million people who saw it hang like a red veil over land and sea with its portentous, threatening, warning quality. The cloud was dust, and the dust was the topsoil of the Middle West, including vast areas of Oklahoma, on its way to be lost in the Atlantic. A combination of monoculture, dust-mulching, a couple of weeks' high wind, had had its inevitable result. The soil of the Middle West was blowing into the Atlantic at a rate which, combined with water-erosion in other parts, could reduce North America into a barren Sahara in a matter of a century.

When, between 1889 and 1900, thousands of farmers were settling Oklahoma, it must have seemed to them that they were founding a new agricultural civilisation which might endure as long as Egypt. The grandsons, and even the sons of these settlers who so swiftly became a disease of their soil, trekked from their ruined farmsteads, their buried or uprooted crops, their dead soil, with the dust of their own making in their eyes and hair, the barren sand of a once fertile plain gritting between their teeth. They went west, to pick fruit in California, in single families, in groups of families, in whole caravans of families, riding in ancient 'jalopies' and everywhere were scowled upon, harried forward lest they became a charge upon some other State. The pitiful procession passed westward, an object of disgust – the *God-dam'd Okies*.

But these *God-dam'd Okies* were the scapegoats of a generation, and the God who had damned them was perhaps after all a Goddess, her name Ceres, Demeter, Maia, or something older and more terrible. And what she damned them for was their corruption, their fundamental ignorance of the Nature of her world, their defiance of the laws of co-operation and return which are the basis of life on this planet.

Corrosion of the Landscape
—————— Herbert Girardet ——————

Like a phoenix, Europe rose out of the ashes of the Second World War. Only a decade later the ruins of the inner cities had given way to gleaming new office blocks and department stores. Everywhere houses were being connected to the electricity grid. Factories were mass-producing washing machines, refrigerators and television sets. Tractors, lorries and cars were being propelled by cheap oil that was flooding in from the Middle East. By the early-1960s Harold Macmillan said to the British people, 'You've never had it so good.'

One of the few new German words to enter the English language was *Wirtschaftswunder*. In West Germany industrial growth was consistently higher than elsewhere in Europe. In the cities, television towers and power-station chimneys grew above the spires of cathedrals on the skyline. Beyond the suburbs, new outer suburbs chewed their way into farmland. And in hidden valleys rubbish tips started to make their appearance, dotted with shreds of plastic.

By the mid-1960s the average family owned a car. Road maps became out of date each year as new motorways stretched their web across the countryside, connecting every city with every other. Trips across Europe in the family car became the norm. The Sunday walk in the woods gave way to the weekend trip to the Adriatic.

The 1970s brought the first shocks of uncertainty represented by oil crises and political kidnappings, the growth of the anti-nuclear movements and the publication of such books as *Limits to Growth*. In Bonn, annual economic growth could no longer be taken for granted but nevertheless a no-growth economy was still considered an aberrant condition, to be avoided at almost any cost. Motorways continued to sprawl across the land but now citizens' action groups made sure that those in power could no longer take people's passive consent for granted. Despoilation and pollution of the countryside, of rivers and air, became major political issues.

In the 1980s, economic stagnation has become a reality, millions are out of work and the future, once again, looks uncertain. Now another new German word is entering the English language: *Waldsterben* – the decline and death of forests. The sick forests of West Germany have received much media coverage abroad, and the question is being asked: Is the *Wirtschaftswunder* the disease of the forests?

136

Sixteenth-century ploughing and sowing, depicted by Simon Benninck (1483–1561)

Top to bottom, *One man and two horses can plough up to one acre per day; One man on a small tractor with a two-furrow plough can turn up to five acres per day; One man on a modern tractor using a four-furrow plough can plough up to fifteen acres per day. Eight-furrow ploughs are now available for large farms, vastly increasing the labour productivity of farmers*

138

Above, *Crop-spraying in Colorado. Each year thousands of farm workers in America become ill from poison sprays;* below, *Grass-harvesting machinery for silage-making. Throughout Europe, such machinery has replaced grazing animals on pastures, and animals are increasingly kept indoors for much of their lives*

Previous page, *Combine harvesters gather chemically-grown grain on a huge, mechanised farm. This land is almost certainly eroding faster than it is being formed from the rock*

Above, *Experiments prove that soil containing worms (left) provides better growing conditions than the compacted soil (right) without earthworms;* below, *Channels dug by earthworms are lined with nutrient-rich mucus – an excellent plant food. In addition, the burrows help the soil to absorb rainwater and thus reduce erosion on hilly land*

142

Previous page, *The Wiltshire plain. The traditional mixture of woodland, hedges and fields is increasingly giving way to grain monocropping;* left, *The remains of a hedge are burnt;* below, *The new pattern of agribusiness*

When making the television series, our team drove down the autobahn, five cars in a row, on our way to the Black Forest. We were all staring at the trees at the side of the motorway, looking for symptoms of the sickness. It was June and the green on the trees was still fresh. In many places the motorway had been cut right through a forest, we could perceive every leaf, every needle of the trees racing past our car. Then we passed a row of trees with a white cross painted on each trunk. One of them also had a sign which read 'I am dying' fastened to the trunk. We could not stop to take a closer look but we noticed these fir trees had thin, transparent crowns. They were not pointed at the top but were flat with straggly branches sticking out in an untidy manner. We whizzed past, 140 kilometres per hour on the speedometer, we had an appointment to keep.

At Freudenstadt, Herr Trefz joined us. He has been a forester for many years and it is now part of his job to show media people around his part of the Black Forest so that they can tell people at home what it looks like. We drove along a winding road half-way up the Hornisgrinde mountain with the purpose of filming the sickness that is now sweeping across this forest, and indeed the forests of Europe.

We got out of the cars by a small lake, surrounded by spruce and fir trees as far as the eye could see. Herr Trefz announced, 'All of the trees around here are sick, every one of them.' We didn't quite believe him at first, but then he explained to us the symptoms of the disease that we knew only from photographs. He showed us the yellow needles, the so-called tinsel syndrome, trees with drooping branches.

It is a distressing experience to walk through a dying forest. Usually the oldest and tallest trees are worst affected. The crowns which one expects to be full and thick with needles or leaves now have a bare look about them, as if some furious storm has ripped them off. These fir and spruce forests which have a reputation for being dark places now let the light through as the trees go bare. Wherever the disease has struck, deformities become apparent. Many trees grow abnormal side-shoots from their trunks. Diseased fir trees often have peculiar flattened tops that look rather like storks' nests. And that is what the foresters call them, though no stork has been seen in these parts for decades.

On many trees the bark was beginning to flake off because of damage by bark beetles and the resin was seeping out. There were plenty of small birds in evidence, taking advantage of the abundant food supply of wood-boring insects.

We climbed a steep track on the side of the mountain, each of us carrying camera equipment and reels of film. Fog came down fast and soon covered the mountain like a thick blanket. And then we reached the place we were looking for: there were dead trees all around us, their bare branches sticking out through the fog. We took our pictures in silence. Then Herr Trefz said,

Two years ago these trees were still green. Now there are thirty hectares of dead trees. How long before all the trees on this slope will have died? We've tried to plant new ones, but even the two-year-old trees have gone yellow. They won't make it through the winter. Acid fog is killing them all.

We used to take pride in our work, planting trees, thinning out young forests, felling trees when they were mature. But it is no longer like that. Our job has changed from forester to undertaker. All we do now, all day, every working day, is to cut down sick and dying trees while their timber is still worth anything, and to prevent the further spread of the bark beetles. Now we work without plan, we simply respond to the situation as we find it. And our colleagues in the Harz, the Fichtelgebirge and part of the Bayrische Wald are in much the same situation. What will it be like in two or three years' time?

The prospects are frightening. In autumn 1982, 8 per cent of forests in Germany were thought to be sick. One year later 34 per cent were considered damaged. In October 1984 the Bonn government announced that estimates by foresters throughout western Germany indicated that 50 per cent of forests were affected. When will it be 100 per cent? And what would that mean to a country one-third of which is covered by forest?

Not every tree that is now considered sick will die within a few years. Not everywhere are things as bad as on the Hornisgrinde, not even in the Black Forest. In many places young trees still seem to be growing vigorously. But the grim truth is that the forests that are most important for environmental protection, that is, mountain forests, are worst affected. In the Alps – in Germany, Austria and Switzerland – the forests are very sick in many places. The first estimates have been drawn up to assess the value of the environmental protection provided by the forests in preventing landslides and avalanches. They suggest that the Swiss government would have to spend £200 billion to build retaining walls and protective barriers against landslides if the protective function of the tree roots, holding soil and rocks in place on the mountain slopes, were to fail. Any damage to farmland, houses, roads, local industries and tourism are not included in this figure.

The suddenness with which the *Waldsterben* is sweeping across Europe has taken everybody by surprise. People still cannot grasp that its relentless progress is so very closely tied up with their everyday lives. It is clear now that the forests are simply unable to cope with the bad breath of industrial society.

The forest disaster is the result of a massive chemical onslaught on the landscape of Europe – and also North America – by the fumes emitted from power-stations, factories, chemical works, refineries and private households. Cars and lorries are also very much responsible.

With hindsight, it is becoming apparent that forests in Europe have been affected by industrial and traffic emissions for up to twenty years. The analysis of tree rings in forests that are now severely affected has shown that tree growth has been impaired since the mid 1960s, when economies expanded ever more rapidly. Emissions of sulphur dioxide from power-station chimneys increased every year until the early 1970s. And output of nitrogen oxides from chimneys and from the exhaust-pipes of cars and lorries is still growing now. Other gases such as hydrogen fluoride and hydrogen chloride have been added to the poisonous cocktail. Their combined effect is causing havoc in the forests.

West Germany is particularly affected as an economically successful country with a proliferation of industries and a high standard of living. Too much industrial density. Too many cars going too fast. Too much restlessness programmed into the 'normal' pattern of everyday life.

The decline of forests is the clearest indication yet of the incompatibility of the industrial growth economy and living nature. Everything nature produces eventually becomes the source of subsistence for another being, nothing is wasted, everything is recirculated. The industrial system on the other hand imposes its emissions relentlessly on the living world. They cannot be beneficially reabsorbed. It is estimated that western Europe releases approximately 25 million tons of sulphur dioxide and 9 million tons of nitrogen oxides into the atmosphere. Since 1950, rain in Europe has become ten to eighty times more acid as a result of these emissions. The tall chimneys that have been built in the last twenty years disperse these gases over a very wide area. Trees, streams, lakes and agricultural crops absorb them in the form of 'dry deposition' and as acid rain, snow and fog. The Black Forest is affected by the emissions of gases from refineries on the French side of the Rhine and from factories and power-stations in the Stuttgart area. The forests of the Harz mountains on the East German border are damaged by polluted air from the Ruhrgebiet, 150 miles to the west. Sweden and Norway are exposed to emission from the Ruhr area and from the Midlands in Britain. The clear, acid, dead lakes of Scandinavia are the best documented example for the chemical onslaught of countries on each other. Steve Elsworth reports in his book *Acid Rain*,

In 1.3 million hectares (5000 square miles) of Norway's southern lakes, the fish are practically extinct and inland stocks have been affected in nearly another 2 million hectares (7500 square miles) – a total area bigger than Belgium. In Sweden, 20 per cent of lakes are affected, and fish in at least 4000 of Canada's lakes have died as a result of acid-carrying pollution. There are also large-scale aquatic deaths in certain parts of the USA, and increasing reports of acidified lakes in the UK.

In Scandinavia, too, investigations into the conditions of forests have revealed damage of considerable proportions, notably in southern Sweden.

This is of particular concern to people in western Europe since a large proportion of our timber imports comes from Scandinavia. But so far there has been no significant response to Scandinavian protests about 'chemical warfare' against their environment by Germany and Britain.

The lakes of Sweden and Norway have been known to suffer from acidification for a decade or more. But the decline of forests all over Europe has been apparent for only a few years. The first reports about sick fir trees in Germany were published in 1979. Three years later, very few of them could still be given a clean bill of health. Spruces, and pines, too, were beginning to show symptoms of distress different from the common diseases that forest pathologists were familiar with. By the autumn of 1983, beeches, hawthorn and ash trees, maples and oaks were all showing signs of decline. Virtually all types of broad-leaved trees were beginning to suffer from new disease symptoms: premature loss of leaves, leaves that did not reach their normal size, brown patches on the edge of leaves. In badly-affected areas, groups of broad-leaved trees were displaying transparent crowns in which many of the smaller branches and twigs had become dry.

Researchers in Germany and elsewhere are still trying to find out why the forest disease has struck so suddenly and with such unexpected ferocity. After all, the calamity began to occur at a time of recession, when emissions of toxic fumes had been stabilised because of a drop in demand for electricity and industrial products. There are many theories about the way in which air pollution affects trees. Some researchers insist that acid rain increases the acidity of forest soil, thereby leaching out beneficial minerals such as magnesium while exposing the roots of the trees to toxic aluminium. This is still accepted as a contributory cause but not the only one. It was found that even trees that grow in soil rich in magnesium and lime show serious symptoms of decline.

Other theories suggest that various infectious organisms – viruses or fungi – are the primary cause of the forest decline, blocking the sap flow from the roots of the tree up to the needles and leaves. But critics argue that this does not explain why so many different species of trees have all been affected at virtually the same time. There is not one single disease organism that is known to affect so many different kinds of trees.

Perhaps the so-called 'stress-hypothesis' put forward by Professor Schütt and his colleagues at the Forestry Department of Munich University offers the most plausible explanation. They suggest that the poisonous cocktail of emissions with which trees have to cope disturbs the process of photosynthesis. Poisoning of the tree by small quantities of toxins over decades reduces its vitality. Its metabolism is impaired and its growth reduced. The weakened tree finds it harder to defend itself against diseases. The reduced function of leaves or needles does not allow the tree to build up necessary reserves of nutrients. It becomes incapable of the vital, regular renewal of the hair-roots. Beneficial fungi associated with the tree roots, which are crucial for the nutrient supply to the tree die off. The brittle roots are attacked by malignant fungi and bacteria, further reducing the capacity of the tree to withstand adverse environmental impact. The tree becomes highly vulnerable to insect damage and to extreme weather conditions such as drought and frosts with which it would be able to cope were it not in a state of constant stress.

But even this hypothesis does not give a clear explanation why symptoms of serious forest decline are being observed all over Europe – and indeed North America – simultaneously. Similar symptoms are now apparent in West Germany, Austria, Switzerland, the Netherlands, Belgium, eastern France, northern Italy and, to a lesser degree, Britain. Czechoslovakia, Poland and East Germany are particularly badly affected with entire hillsides near industrial centres denuded of trees. The Erzgebirge in Czechoslovakia is the most shocking example of all: around 245,000 acres (100,000 hectares) of forests have died where a decade ago the trees still appeared to be thriving. In the Soviet Union, too, major damage to forests has now been reported.

It has been suggested that damage in Eastern Europe is caused primarily by local sulphur dioxide pollution of the sort which has been observed since the industrial revolution. The picture in western Europe seems much more complicated. The emission of a great variety of poisonous gases associated with our ever more complex industrial processes affects trees in ways which have not been observed before. The toxicity of one gas reinforces the impact of another gas. Sulphur dioxide has a major impact on long-lived plant organisms in the winter months, while nitrogen oxides have their worst effect on hot sunny days when chemical reactions under the influence of sunlight result in the release of huge quantities of corrosive ozone.

One irony of the forest disaster is the temporary glut of timber that is beginning to flood the European market. Since foresters prefer to fell diseased trees before their trunks are riddled with wood-boring insects, large quantities of timber are now coming on the market. In West Germany the possibility of special timber depots is already under discussion in which felled trees which are surplus to immediate market requirements could be stored. A West German government institute for forestry research has already investigated how much timber would have to be deposited if all conifers older than fifty years had to be felled in the next fifteen years. The total quantity would be 6000 million cubic yards, enough to cover the territory of the county of Hamburg with a pile of timber one yard high.

Researchers are now becoming increasingly concerned about the impact of industrial, power-station and traffic emissions on the *soil*. It is a long-established fact that worms are affected by soil acidification; there are fewer worms in acid soil. A considerable reduction of soil life in its great variety has been observed in forests affected by acid rain and industrial pollution. Soil compaction and the reduced ability of the soil to absorb water are inevitable results.

Professor Preuschen, for many years director of the Max-Planck Institute for Agricultural Work and Technology, run by the West German government, is greatly concerned about the impact of airborne pollution on agricultural soils. His research has shown that life in the soil, already impaired by modern agricultural methods, is being further damaged by acidification. Of course, agricultural soils can be limed to counteract acidity and this is being done now at an average rate of 50 lbs per acre (60 kilograms per hectare). But as the rain has become increasingly acid – the pH has fallen in the last twenty years from 5.2 to 4.1 – soils are constantly exposed to renewed acidification. Preuschen is greatly concerned about the effect this is beginning to have on crops. In the Nürnberg area, he has observed the condition of grain fields for the last three years and he found that barley, in particular, had a pale-yellow colouring instead of the normal dark-green. Growth was clearly impaired. Preuschen came to the conclusion that soil life in the barley fields was severely affected by sulphuric acid and nitric acid which comes down with the rain. He is convinced that it cannot be long before yields are affected. His views are shared by other soil scientists in Germany.

Research in Britain has shown that acid rain is causing damage to bean crops. A recent article in the magazine *Nature* states that the chemistry of bean plants is changed by acidic rainfall, making them more prone to aphid attack and requiring greater dosages of insecticides as a consequence. These conclusions were reached by researchers investigating the increasing abundance of aphids on crops in south-eastern England, downwind of London.

Professor Preuschen and several of his colleagues are alarmed at the prospects for European agriculture if acid rainfall continues at the present rate. Not only are soil micro-organisms damaged and killed, but crops are also directly affected. Acidified pastures have a much reduced variety of grasses and herbs. Fruit trees in Baden-Württemberg have started to show symptoms of decline similar to those observed on forest trees.

There is also increasing concern about the accumulation of heavy metals in the soil, notably cadmium and lead. Cadmium reaches the soil via industrial emissions and also through the application of phosphate fertiliser. As much as 35 per cent of the build-up of cadmium in the soil in West

Germany is traced back to phosphate in which it is contained as an impurity. Cadmium in the soil is mobilised by acid rain. It is absorbed by crops and also finds its way into the groundwater.

A Berlin soil scientist, Professor Kloke, has recently estimated that as much as 7 *per cent* of the surface area of West Germany is so badly affected by heavy metals, including cadmium and lead, that the soil should no longer be used for growing crops for human consumption. The build-up of cadmium in the human body leads to damage of the lungs, kidneys and bones. The environmental organisation BUND has proclaimed that if heavy metal accumulation in the soil continues at present rates for the next fifty years West Germany will no longer be able to grow *any* crops fit for human consumption.

In Britain, too, high heavy metal concentrations have been observed in city areas and near motorways. In some cities in the Midlands, allotment growers are being warned that their crops may not be fit to eat. What will further research reveal?

In Germany, the threat to soils from airborne pollution and present agricultural methods is being taken seriously by central and local government. In 1979, a group of eminent scientists was commissioned by the Bonn government to work out an ecological action programme for the 1980s. They published a 200-page report in 1983 which emphasised the great importance of soil protection. The report stressed that traffic and industrial emissions must be reduced; that agricultural chemicals, particularly pesticides, are doing damage to soils; that the wholesale use of pesticides should be abandoned in favour of 'integrated pest management' in which pesticides should be applied only when pests pose a serious economic threat to the farmer. A transition to biological agriculture was recommended though it was argued that this must be a long-term goal. The report stated that 'ecology is long-term economy' and that nature as the basis of our well-being must be considered at all stages of economic planning.

The scientists recommended that the duty to use soil in an ecologically suitable manner should become enshrined in the constitution. So far, any action by the Bonn government in response to the report has been shelved, presumably because of its potential impact on well-established economic interests. However, the government of the Land Nordrhein-Westfalen, the most populous region of West Germany, has now committed itself to enact a policy of soil protection and ecologically acceptable agriculture which will have far-reaching consequences. The programme emphasises the need to protect soil from toxic emissions. It calls for research on suitable methods for applying sewage and animal wastes to farmland. It stresses that the impact of mineral fertilisers on groundwater must be reduced and that suitable cultivation methods must be practised to prevent soil erosion. The use of pesticides should be minimised in favour of suitable techniques of integrated pest management. The agriculture minister of Nordrhein-Westfalen, Herr Matthiesen, said in a press conference to publicise the programme that he could see no alternative to the creation of a system of ecologically-viable agricultural techniques. There are 100,000 farmers in the highly industrialised part of Germany and it will be interesting to see how the state government will induce them to adopt this high-minded programme. And how will the protection of the soil from industrial emissions be enforced?

West Germany is the most intensely industrialised country in Europe, perhaps in the world. The condition of the soil in that country which is now causing such concern must be food for thought for industrial planners in countries trying to emulate the success of Germany. The blessings of a way of life which is so greatly dependent on industrial throughput are now seriously in question. The deplorable condition of the forests has sent shock waves through the population and many people are beginning to wonder whether the post-war industrial *tour de force* has really been worthwhile. A period of great affluence for one generation cannot, must not, be bought at the expense of the living chances of future generations.

Cancerous fish in rivers such as the Elbe and the Weser, dying forests, polluted soil, groundwater and air – are these the necessary consequences of a *Wirtschaftswunder*? We would all like to think otherwise since we are all involved to a greater or lesser extent.

The now rampant *Waldsterben* has forced the West German government to change its previously complacent attitude to air pollution. It has started to enact measures which will reduce emissions of sulphur dioxide and nitrogen oxide from large power-stations by around 50 per cent within ten years. It has persuaded other European governments to take similar measures in their countries. West Germany is also determined to reduce nitrogen oxide emissions from car exhausts and is pressing ahead with plans to make catalytic converters compulsory in the exhaust-systems of new cars. This will cause a rapid transition to the use of lead-free petrol since the converters which clean nitrogen oxide out of the exhaust gases cease to function with leaded petrol.

Many people fear that the new measures are a case of too little, too late. The available evidence suggests that a massive reduction of pollutants from the atmosphere is necessary *now* if the forests of Germany and neighbouring countries are to be saved. Gerhard Weiser, agriculture and forestry minister for Baden-Württemberg, says that if pollution continues at the present rate most of the fir and spruce trees will be dead by the 1990s. It is clear that the damage will be overwhelming by the time measures to clean up the air – including the toxic gases from power-stations and refineries in the Alsace which are affecting the Black Forest – prove to be effective. In West Germany, the state of the environment is fast becoming people's greatest worry. In an opinion poll conducted by the EEC in 1982, 77 per cent of those who participated said that the condition of the environment troubled them more than unemployment, worsening international tensions or increasing crime rates. This sentiment is making its mark, setting new priorities for the political agenda of the remainder of this century. The success of the Greens has shaken the established parties. A politician who does not have strong opinions on how to improve the state of the environment now has little chance of being elected.

Few establishment politicians would accept that industrial growth economics is the root cause of environmental degradation. They point out that Japan has the fastest growing economy of them all and has succeeded in tackling pollution. Since the late 1960s, successive Japanese governments have taken active measures to reduce sulphur dioxide output from factories and power-stations. Up to 90 per cent of sulphur emissions are now intercepted before they reach the atmosphere. Cars are fitted with catalytic converters to reduce nitrogen oxide in the exhaust gases. Nitrogen oxide emissions from power-stations are also being reduced now under a government programme introduced in 1981. Japan will be in a good position to sell its pollution control technology to European countries when they begin to tackle the problem.

But in Japan, too, the trees are sick. For almost a decade now, pine trees have been affected by a virulent disease, pine blight. In an article in *Resurgence*, Alfred Quarto described the decline of the pine trees as a problem which has reached such proportions that it is a national disaster:

I can vividly recall that odd sight I witnessed while visiting one of the ancient temple gardens in Kyoto. In this particular garden an old and once beautiful pine stood, but it was wrapped from base to tree-top, including all its graceful branches, in white bandages. A bottle containing a clear liquid was strapped to its side, and a feeding line ran beneath the bark. The tree's caretaker told me this tree would probably not live.

A well-known expert and teacher of organic farming, Masanobu Fukuoka, who has studied pine blight for several years, has come to the conclusion that the crucial symbiotic association between the roots of the pine tree and a certain fungus is impaired by industrial emissions. The trees are no longer able to defend themselves against diseases which would normally be relatively harmless. Fukuoka's findings are still being disputed by government scientists but they have not yet produced plausible explanations for the plight of the pines. The Japanese are quite fanatical about their trees – nearly 70 per cent of the country is covered by forests – and there is great public concern about the future of the pines.

Living organisms can only resist so much pollution. The problem is that we do not know exactly how much. Since the war we have been involved in a gigantic random experiment to find out. In the northern hemisphere we have subjected millions of square miles of forests and farmland to varying levels of air pollution. Now the results of our tests are starting to come in.

Our biggest problem is that we did not quite realise that trees are not as disposable as laboratory mice. And neither is soil life. We need the many varied organisms of the natural world for our survival. Future generations need earthworms in the soil in which they will grow their crops as much as we do today. We also need technology. But who said: 'Technology is the answer; but what was the question'?

Industrial growth economics and technological innovation for its own sake are being viewed with increasing scepticism by people throughout Europe. Opinion polls show that more and more people are willing to forgo the benefits of a higher standard of living for the sake of a healthy environment.

Modern industrial technology has made the medieval dream of the perpetuum mobile come true. But only now are we beginning to realise that this seemingly perpetual motion is dependent on burning finite fossil fuels and that it causes pollution which is not compatible with the essential requirements of living organisms like trees. Unfortunately, trees cannot walk away if the air doesn't suit them. They cannot take off for a holiday in cleaner air.

Freudenstadt in the Black Forest is a well-known holiday resort frequented by people from the cities who want a change of air. The visitors continue to come in large numbers to go for walks in the forest which stretches for many miles in all directions. But it is becoming difficult for people to take pretty holiday snaps. There are still trees everywhere but they are sick and do not make good backdrops for portraits. For several years, the owners of hotels and restaurants were fiercely opposed to environmental groups mounting public protests against the decline of the trees. Recently there has been a change of heart. The *Waldsterben* is there for all to see, the local people have united in wanting to draw attention to it. There is money at stake, for they fear what will happen if the tourists stay away. Since the locals feel that their future is in question, they are now prepared to support ecological action.

A forest on the edge of town has been dedicated to the members of parliament in Bonn. Every MP has been given his or her own tree with a name tag on it. Regular reports are sent to the Bundestag to inform MPs how their trees are doing. So far there are no signs of recovery.

154

A Look at Kenya
— John Seymour —

A man with a King's African Rifles hat with a feather stuck in it blows a whistle – one of those whistles that have a pea in them to make them even shriller – as he does a wild sort of a dance. The noise of the whistle is by no means drowned by four men and boys who are bashing a series of drums. About a dozen people, men and women dressed or undressed for the occasion, perform a furious dance. All this is taking place at a little village of Kyevaluki, near Kangundo, near Machakos, which is just west-south-west of Nairobi, in Kenya. The noise is mighty, the sun is roastingly hot, a lot of dust is kicked up by the dancers and it is altogether a very spirited scene. The dancers are also watched by some Christian people, mostly, like the dancers, of the waKamba tribe, but much more soberly dressed and modestly behaved, and they are watched also by two missionaries, one an Englishman and one a Welshman. These men are strict Baptists and look on the scene with some disfavour. They have tried praying for rain, in a much more restrained manner no doubt, but to no avail. What if these pagan rain dancers prevail?

A very undersized goat is dragged out, crying piteously, its throat is cut and the blood poured into a calabash. The blood is sprinkled on to the ground as prayers are said, or chanted. Some of the members of the camera crew blanch slightly. The missionaries avert their faces.

We go into a small rectangular hut with a corrugated iron roof. It is stifling in there. The hut is divided into two compartments: one is just big enough to hold a double bed, the other a small table with some boxes around it to sit on. We are lavishly fed on pancakes. I reflect that the people who are giving us the pancakes can ill spare the flour – they are heading towards starvation themselves.

There are several other huts in the small hamlet, which is the home of one extended family. Some of the huts are rectangular, some are round. The rectangular huts all have iron roofs and are like ovens inside. I look into one of the round huts, which has walls and a roof of grass, and find it deliciously cool.

I remember the Africa I knew before the war. I never saw a rectangular hut then with an African family in it. I have slept, many scores of nights, and sweated out many scores of days, in thatched African huts. But I did not sweat too hard. These huts were fairly cool during the day, when it was roasting outside, and very warm at night. In the evening there would be a little smouldering fire in the middle of the earth floor. When you first walked into a hut the smoke was choking but, when you knew the trick, you immediately sat down. On the floor, of course, or on a reed mat or a skin – there were no chairs. And then your head came below the smoke. There were no mosquitoes as the smoke kept them away. The comparative dark, during daytime, kept the flies away, too.

The hut in which we ate the pancakes belonged to a young man named Philip Kiundu. I asked him why he lived in a tin hut when a grass one was so much more comfortable and I gathered from his answer that it was for reasons of prestige. He fully admitted that it was not so comfortable and he preferred to sit in his mother's hut, which was old-fashioned.

There was a *Mzee* (old man) with a King's African Rifles hat on and I found that he had been an *askari* (soldier) in the same brigade as my own. We had both been in the Gondar campaign, in northern Ethiopia and in Burma. I must have seen the man many times. We sat in the corner of Philip's hut and chatted away, in Swahili, of old times. I found that, after forty years' disuse, my Swahili came welling back. He deplored the fact that the young people of the waKamba tribe were losing respect for their elders and their ancient traditions. All the young men could only think of one thing – money. They wanted to have prestigious tin roofs for example, although it was possible to build a traditional African hut in a day which would cost exactly nothing.

So the young men streamed into Nairobi and tried to get jobs. Mostly they could not and turned to robbing people or begging. All the parents thought that if their children could only pass school examinations they could get jobs in Nairobi as clerks and be happy ever after. But who would feed them? As for this *Mzee* he could neither read nor write, and knew no English, but he farmed his *shamba* (small farm) well, his cows multiplied and were healthy and his crops were good. I asked him about soil erosion. He was well aware of it and said that if only the young men would stay at home they could plant trees, terrace their hillsides and stop erosion altogether. But the farming was done by old men and women: the young men looked at work on the land as being beneath them.

Young Philip told us of the great sacrifices people had to make to send their children to school. At the local school the parents had to pay £100 a year in fees. They had to pay a third of this in advance before every school term. Three times a year, the price of cows plunged disastrously, because so many parents had to sell their cows to raise the school fees. A heifer or young bull would fetch only £35, an older beast perhaps £70.

People had to flog their land to get the money to pay these school fees. They had to grow maize on slopes where no maize should grow, because of erosion.

To see the fruits of all this sacrifice we repaired to the Mulli School in a village some miles away. There I talked to some of the children.

Their home language was kiKamba but they spoke Swahili, the lingua franca of East Africa, very well. They all knew some English but it was rudimentary. I gathered from talking to them that they understood very little of what they were being taught at school.

Their headmaster, who was the English Baptist, spoke no Swahili and no kiKamba, but intended to learn the latter. He gave the assembled children a speech, talking to them in the sort of English that would have gone right over the heads of a class of English children of the same age, and it is my opinion that they did not understand a word of it.

The headmaster told me afterwards that the children were enormously keen to achieve high marks in their examinations because they knew of the great sacrifices their parents were making to send them to school.

I wandered into an empty classroom while all this was going on and looked at a textbook lying there. It was a history of Russia, pre-Revolution. I reflected on the relevance that this could have for the children of that school, in the impossible event that they could understand it at all. I felt like going to the headmaster and saying to him, 'Throw that silly book away and take those children up on to the hillsides and teach them how to build terraces! Teach them how to conserve and use the dung of their cattle! Teach them how to avoid overgrazing and tramping-out of pasture! Teach them to save their *land*!' But I did not.

It did not rain, and I *imagined* I detected a certain relief on the part of the Christians.

We were taken out into the uKamba country again by a splendid Dutchman named Willem Beets, who worked for ICRAF, which stands for the International Council for Research into Agro-Forestry.

They are trying to persuade African farmers to plant trees among the crops on their arable land. We saw one splendid example of their success: a substantial farmer, by local standards, who had wholeheartedly adopted the ICRAF suggestions.

His land was on a slightly sloping plain to the east of a steep-sided mountain range. The range had, alas, been denuded of its original tree cover by charcoal burners: there is a high price to be obtained for charcoal in Nairobi and, although the government tries to stop them, with so much money at stake *nothing* can stop them.

The result of the removal of the trees means that the streams and rivers running from the mountains are no longer gentle and clear but are often empty when it is not raining, and gorged with fierce, eroding floods when it is. Owing to the pressure of population, and the need for everybody to earn money to educate their children and to buy tin for the roofs of their houses, cultivators are working higher and higher up on the lower slopes of the mountains, with results only too easy to observe. Fierce gully erosion is taking place.

Down on the plain, on most of the *shambas*, the rule is more generally sheet erosion, with occasional gulleys developing here and there. But the soil is *going*, there is no doubt whatever about that. The topsoil has already gone and only artificial fertilisers will grow crops now. To pay for these the farmers must exploit their soils even more.

But on Mr Japhet's farm things are different. The soil is not going at all. All along the fields, whether they be growing maize (which most of them were) or other crops, there were rows of small trees. These mostly belonged to the family *leguminosae*, the pea-and-bean family, and had been chosen for their ability to fix nitrogen from the air, thus relieving Mr Japhet of the need to buy fixed nitrogen from the co-operative.

The trees never got beyond a certain size, because Mr Japhet cut their branches from time to time to feed to his cattle. The latter spent some time grazing the nearby hillside, in charge of a small boy, but also some time confined in a *kraal*, or *boma*, a thorn-bush pen. Their manure accumulated there and, when Mr Japhet had the labour, he carted it out on to his land. Unfortunately this is seldom done thereabouts and the manure is simply wasted because it just wastes away.

Obviously some land is lost to cash crops because of the room the trees take up but this is, in the long run, more than compensated for by the fact that the trees stop wind desiccation of plants or soil, stop soil erosion and add humus to the soil by their leaf-fall. The bonus of fodder for the cattle is extremely valuable as this helps to spare the sparse grazing. The resultant manure is made available to return to the land. Mr Japhet told us that he had doubts about the system for the first two or three years when his crop production fell, but now his yield is increasing steadily and he grows far more per year than his neighbours on the same amount of land. The trees also supply him with sufficient firewood for his needs.

He has a large orchard of food-bearing trees. Walking in from the blinding dry heat of the drought-stricken country round about into the shade and coolness of Mr Japhet's farm was a delightful contrast.

Mr Japhet told us that before the hill was denuded of trees there was far more water to dispose of for irrigation as the streams ran for most of the year. Now they ran not at all except for occasional wild rushes when it rained hard. It occurred to me that if every Kenyan farmer had Mr Japhet's attitude, if the charcoal burning and other deforestation would stop, if trees were again planted on the uplands, if the manure from the cattle were once more spread out upon the land, if cattle and goat and sheep grazing were organised sensibly, in a controlled way, with areas of the bush being rested so they could recuperate ... Kenya could become the paradise it was when the white men first came to it in the late nineteenth century.

A few days later, we witnessed an interesting sight: a group of about forty women who had gathered together to stop a *donga*, or erosion gulley, which had recently developed.

This *donga*-filling is becoming quite a popular activity in Kenya. Even the President, Mr Moi, goes out into the country somewhere on most Sundays to have his photograph taken throwing rocks into a *donga*.

The women (nearly all 'manual' work in the countryside in Kenya is done by women) sing as they work to a kind of dance rhythm. The dance rhythm obviously slows down the work a little but it makes the work more enjoyable. Two young American girls are with us: they are working for ICRA. They both express doubts as to whether the work being done is going to be of the slightest good. The torrential floods, in the next rain, will simply work round the ends of the bank of stone and the whole thing will crumble and continue to cut into the precious soil of Kenya. They are in favour of planting trees all around the *donga*, to contain it that way. But who would stop the goats from killing the young trees?

They take us to see an effort, run by local people under the guidance of ICRAF, to establish trees. It is a tree nursery, situated by a little dam that actually has water in it. Women are sieving earth to plant seedlings in, transplanting seedlings and watering them with water from the dam. Again, not one man is at work. The old tradition is that the men hunted, fought off other tribes and herded cattle while the women did all cultivation work. Now that there is nothing left to hunt, the central government keeps peace among the tribes, and pressure of population has made agriculture far more important than pastoralism, nine-tenths of the work on the countryside devolves on the women. Most of them have to carry water from long distances, firewood from even further, bear and look after the children, brew the beer and cook for the men. If it is true that Satan finds work for idle hands to do the women of Africa are quite safe. As for the men – a clerkhood in Nairobi is all their ambition. If they achieve that – what does it matter what happens to the soil back on the *shamba*? There are, I must add, honourable exceptions.

During the war I spent many months in what we used to call then the NFD – the Northern Frontier District. The whole of the northern part of Kenya is either desert or semi-desert, and the best that can be said about it is that some of it is 'sparse bush' or 'sparse savannah'.

All around the very dry area called the Chalbi Desert I noticed that there were large piles of stones scattered about, quite obviously assembled by man. One day, on leave in Nairobi, I went to the Coryndon Museum and asked the curator, who happened to be none other than the celebrated

Professor Leakey (who later discovered the earliest traces of man on this planet) what they were. He said that without a doubt they were burial mounds of ancient people. That land, he explained, had once had soil on it and been well inhabited.

Now it is a howling wilderness. The Chalbi itself is sand but the surrounding country for hundreds of miles is bare lava rock, with little thorn bushes struggling to survive, scant grass after the rains, and the only inhabitants are nomadic Boran people who, in those days, lived off great herds of camels.

Leakey said that he did not think the soil had gone because the climate had changed. The rainfall is still quite adequate. He said that it had gone because it had been abused by people. If anybody who farms land in a hot dry climate wishes to see what his farm will eventually look like if he does not farm it sensibly he should go and look at the countryside around the Chalbi Desert.

So we revisited this country to see what is happening there now to the Boran people.

We flew from Nairobi in a light aeroplane to a landing strip at a place called Malka Dakaa. The landing strip had not been used for many moons and the pilot expressed great satisfaction that he had managed to avoid a nasty hole which the recent rains had created, otherwise we would not have had a very happy landing. He taxied his aeroplane under the shade of a big thorn tree where he would spend all day waiting for us.

The country here, much further west than the Chalbi, has more and bigger thorn trees and is obviously more productive. The heat, although it was still early in the morning, bounced back from the bare ground and we were soon assailed by the ferocious thirst that I remembered so well. A large Land Rover was waiting for us at the airfield. The fact that it had several bullet holes in its body did not fill us with confidence and I think we had mixed feelings about the two *askaris*, armed with automatic rifles, who perched on the back of the vehicle and were with us, obviously quite alert for possible trouble, all the while. There was a little war between the Somalis, from the east, and the Boran, and there had been much bloodshed. The driver, and owner, of the Land Rover, apologised for the bullet holes and said he did get shot at occasionally but the *shifta*, the Somali guerrillas, were rotten shots, so we all crammed in.

He drove us to the village of Malka Dakaa. The village was a collection of huts, some of them the old round Boran huts that I remembered from the war days, but most of them were little rectangular ovens of corrugated iron. The Boran are extremely *beautiful* people. The young girls have a gazelle-like grace, and the extremely hard lives they lead in that wilderness of northern Kenya keeps them slim and lithe. They have the thin refined

Intensive animal farming: Above left, *Calves bred for veal;* above right, *Battery hen-keeping;* below, *Caged sows*

Above, Evening View of Cyfarthfa Works, *1896, by T. Prytherch – the furnaces and smoke stacks of the Industrial Revolution;* below, *The lignite mine at Hambach, West Germany*

Above, *Port Talbot steelworks in Wales — industrial emissions are now thought responsible for the death of forests;* below, *Broadleaved woods in early autumn*

Opposite, *Dying spruce trees in Germany, affected by industrial emissions from the Ruhrgebiet;* above, *The explosion at the Union-Carbide plant in 1984 at Bhopal, India, killed over 2500 people;* below, *Construction of a pipeline for pollution-free water in Rheinhessen, West Germany*

features of the Hamitic race, to which they belong, and longish black hair.

The Boran in this area were cattle people, not camel herders. But they had very few cattle. The reason for this is that Somali raiders had taken most of them. The Somali people have never recognised the suzerainty of Kenya over the North Eastern Province, which has traditionally been Somali grazing grounds.

We had come there to see two things, one a failure and the other a success. The failure was a grandiose irrigation scheme financed by FAO and UNDP. We saw the land from which the bush had been cleared for agriculture, all but an acre or two of it reverting to bush again; we saw the uncompleted canal which was to lead water from the Ewaso Nyiro River to the fields; and we saw a huge machinery park, full of machinery. We calculated that there was at least half a million pounds worth of machinery lying there. There were huge tractors, huge scrapers, huge bottom-dumpers and a great variety of other heavy plant. Not one machine was in working order except one small tractor, which was being used for giving children rides up and down the village street. The tractors, particularly, had been pillaged for spare parts to be sold to other parts of the country. It had cost vast sums to get all this junk to the place and the cost of getting any of it back again was not considered worthwhile.

We had with us a young man named Richard Hogg, an anthropologist, who spoke the Boran language and had made a study of the people. He had been employed by Oxfam, and partly financed by the World Food Prog-ramme, to try to alleviate the frightful misery of the people of this area. Richard's idea was to turn the people into pastoralists again.

Unusually, the pasture around this area is undergrazed. This is because of the loss of most of the grazing animals to the Somalis. Richard Hogg decided to use the Oxfam money to buy sheep and goats, and some donkeys for transport, to fit out seventy of the most deserving, but capable, families as happy nomads again. Each family was given five milk-tooth doe goats, three bucks, thirty-five milk-tooth ewe sheep, one ram and six wethers. Donkeys proved to be unobtainable at a reasonable price. Cattle were considered too vulnerable to raiders. Besides the stock each family received a *panga* (small sword) and two plastic water-containers. In places like Malka Dakaa things such as plastic water-containers are hard to acquire. We went out into the bush and saw some of the sheep and goats with their happy owners.

Previous page, *Cancerous fish from the heavily-polluted Elbe estuary in West Germany;* inset, *Oxygen tanks on the shores of the Baldecker See in Switzerland – oxygen is pumped into the lake to revive the water which has been eutrophied due to slurry and sewage run-off from the surrounding farmland;* opposite, *The Ethiopian famine – hunger is a problem of climate but is also one of unequal food distribution and use of good farmland to export cash crops*

The local Catholic missionaries, incidentally, are going to salvage some of the irrigation scheme, using their own and local hand labour instead of huge machines and, from what I have seen of their work in other parts of Africa, I think they will succeed.

What conclusions were we to draw from what we saw? I think the chief lesson is 'small is beautiful', in Africa as everywhere else. Looking through some old files I found an article from 1947 called 'A Plan for the Mechanised Production of Groundnuts in East and Central Africa'. This scheme, dreamed up by the post-war British government, aimed to 'mobilise 3,250,000 acres in Tanganyika, Northern Rhodesia and Kenya. The land will be farmed in units of 30,000 acres, each employing seven Europeans and 300 natives. Mechanisation . . . will be, in the words of a Rhodesian correspondent, "as near as dammit 100 per cent".'

The initial capital expenditure was £24 million but before the scheme was abandoned an incomparably larger sum than that had gone down the drain and, far more seriously, a vast area of African bush country had been ravaged, the game destroyed, the trees bulldozed and burnt, and the soil itself exposed to terminal erosion. *Nothing* except chemicals was to be put back into this land. As one of the advisers, Dr Martin Leake, formerly of Rothamstead – the English agricultural research station, wrote, 'The humus factor, though recognised, is in fact relegated to a purely secondary role and the problem of soil fertility is considered from a purely chemical angle of replacing by fertilisers the plant nutrients removed.'

Exactly. In a groundnut shell. The 'inorganic approach'. The simplistic approach which is bidding fair to ruin our planet. The approach that can only take account of things that can be *measured*.

Mechanisation 'as near as dammit 100 per cent'! But what is the sense of mechanisation in a country that has, above everything else, a huge surplus of labour? Ordinary African tribesmen had been growing groundnuts, with complete success, in that country for 100 years. They had done it without destroying all the wild animals, without raping the bush, and with the aid of nothing more complicated than a hoe.

After the war I returned to Africa to write a book about it, *One Man's Africa*, and I was taken up by the Kenyan government to Kikuyuland, which was in the throes of the Mau Mau rebellion. I saw some good things and I saw some bad things. I saw quite spectacular anti-erosion works going on: terracing, paddocking for livestock, tree planting, cut-off drains and stone-paved spillways to get rid of surplus water without damage, and an attempt made to persuade the cultivators to return manure to their land and grow green manure crops and fodder crops for their cattle. All good sound organic methods of husbandry which would save the land. I also saw the results of the 'villagisation' scheme of the colonial government,

which was an attempt to force the scattered Kikuyu people into large
fortified villages, so as to deny the Mau Mau insurgents food and sustenance. The villagisation struck me as being unnatural and completely
wrong. But one good thing that was happening was that a few cash crops
were being introduced into the countryside.

The only hope of Africa lies in subsistence farming – people growing
their own food on their own land. But, because of outside influences, this is
no longer enough. To stop people drifting to the towns there must also be
some cash. In pre-independence Kenya, Africans were not allowed to grow
coffee. It was considered that they would lower the standard of Kenya
coffee and depress its price on the world market. Well they are growing
coffee perfectly satisfactorily now. But what I saw in Kikuyuland was the
growing of pineapples. All the peasants were growing these, mostly on
terraced land, very well, and it seemed to me to be a most excellent thing.
Every cultivator could set aside a little of the land on his *shamba* to grow a
crop that would give him some cash, which would improve life somewhat
for himself and his family. Further, he could rotate the land that the
pineapples were grown on to minimise disease and soil-exhaustion problems, and he could also return the manure from his cattle to the land where
it belonged. Small, it seemed to me, was very beautiful.

However, the emergent Kenya politicians, after independence, were
looking frantically for anything that could bring foreign exchange into the
country. Didn't they need skyscrapers in Nairobi? Don't all politicians
need large American cars? And petrol to fill their tanks with? They do.

So Del Monte arrived and persuaded the Kenya government to let them
buy out the existing canning factory – with the condition that, in addition
to perhaps growing some pineapples themselves, they should continue
buying from smallholders too. So they moved in, with vast imports of
American-made machines and, within three years, in 1968 in fact, they
decided that independent growers had to be 'phased out'. Phased out they
were, and they suffered great hardship during this operation. The company now grows pineapples by monoculture on 10,000 acres (4050 hectares) and produces, annually, 60,000 tons of canned fruit. An enterprise
of this size obviously employs a lot of labour, but it employs a fraction of
the number of people who used to benefit by growing fruit on their
shambas. The 10,000 acres the company's farm occupies could, at the
current size of smallholdings in Kenya, support 10,000 families and grow
quantities of good food in a hungry country instead of luxury canned fruit
for consumption by wealthy Europeans. But such an arrangement would
not pay for any more Nairobi skyscrapers, or provide petrol for the cars of
politicians, nor would it fill the pockets of American shareholders.

I had the impression that it is the dead weight of the big cities that is ruining the soil of Africa. To support millions of idlers in the capital cities: Nairobi, Lusaka, Harare, Addis Ababa, the cultivators – and chiefly the women – have an intolerable burden. Education is entirely city-orientated. The children are taught nothing that fits them for the farming life or that makes them feel that farming is an honourable occupation. The thrust of education in Africa is to make children dissatisfied with the country life. So huge export agricultural industries have to be set up in these countries, to grow luxuries for the wealthy nations to exchange for city goods. Fortunately for Kenya the chief export crops are coffee and tea, both of which crops, if properly grown, do not damage the land too much. The pernicious custom of keeping tea land clean-hoed between the bushes on steep land (which I saw to my horror in Sri Lanka during and after the war), is not practised.

We visited a tea estate not far from Nairobi. The estate was owned by a wealthy Kenyan but managed by an Englishman and it was well managed. When, after Independence in 1963, the big British-owned estates were taken over for Africanisation, most of them were retained as big estates and allotted to rich Kenyans, mostly politicians. Some, however, were divided up among the *Wananchi*, the poorer peasants and this did, to some extent, relieve the land shortage. Also the Kenyan government has resisted the attempts of the big estates and multinational companies to expand their acreages at the expense of smallholders. The Kenya Planters' Co-operative Union and the Kenya Tea Development Authority have done good work in helping and encouraging small growers – for example, of the 180,000 acres (72,000 hectares) of planted tea, smallholders have 125,000 acres (50,000 hectares) of them. Most of the balance is owned by big foreign companies.

The population of Kenya is expanding at the rate of 4 per cent per year, which is practically a world record. The Minister of Agriculture reckons that by 1989 another 2 million acres of land will have to be brought under cultivation to provide food for the increased population. Already, Kenya has to import much food. The 2 million acres just cannot be found, or could only be found by reducing the export crop acreage. So whatever happens, if the population rise continues, Kenyans will starve. What we see now in Ethiopia we shall see again further south.

Erosion of soil and soil degradation is galloping in Kenya. One receives the impression that it could get entirely out of control. The Minister of Agriculture and the President are both very well aware of it and are doing whatever they can to arrest the damage. Good things *are* happening. There is some reafforestation (but it cannot possibly keep pace with the charcoal burners), some effort to dam gulleys and, above all, much good terracing. There is one kind of terracing called *fanya juu*, greatly encouraged by the government and other advisers. *Fanya juu* means to make upwards. People shovel lines of earth upwards, along the contours, to make contour banks. Napier grass, and sometimes (all too rarely) small trees, are planted on these earth banks. Land thus treated hardly erodes at all – provided the system is maintained.

Obviously, there is a limit to the number of people who can subsist on any bit of earth. I believe that many more people than now could live in Kenya provided that Kenyan men could be inspired, somehow, to return to their heritage, the land, and husband it properly.

How Far from Paradise?
——— Herbert Girardet ———

Mount Kilimanjaro rises out of the hot dry savannah of northern Tanzania like a gigantic green cone, its summit topped with a cap of ice. When a missionary in 1848 first reported back to headquarters in Europe that he had found a mountain just south of the equator which was crowned with a glacier he was greeted with plain disbelief.

We got out of our beds in Moshi before sunrise, for only in the early morning can one see the great mountain without its veil of clouds. The ice sparkles brilliantly in the early sun; from the ice-cap the water starts to trickle down the mountainside in ancient stream-beds that wind their way through the moors and the forest. Lower down, where the mountain slopes meet up with the plain, live the Chagga in their tree-covered home gardens; it was these we had come to film.

The Chagga are Bantu speakers who originated from various tribal groups some of which, like the Masai, had been cattle herders in the plains below. They moved into the lush rain forests on the southern slopes of the mountain several hundred years ago. They felled some of the trees but left many standing and under the open canopy they planted banana trees. Today, the Chagga gardens attract a growing number of 'scientific visitors' because here is a system of cultivation which is of great relevance for hilly areas throughout the tropics.

We drove into the 'food forest' on a winding track and got out of our Land Rover. We were in a world of green, surrounded by leaves of a great variety of different shapes. High above us we could see the sky through the crowns of the big trees. All around us were banana trees laden with bunches of green bananas. Below and between them grew coffee bushes and at ground level there were vegetables in abundance. We were in a man-made jungle, with houses, huts and animal shelters scattered among the trees and bushes. Everywhere food was being grown, obviously with great success. The Chagga gardens or *shambas* support the highest population density anywhere in Tanzania: 1300 people per square mile, (500 people per square kilometre). Over half the coffee grown in that country comes from the slopes of Kilimanjaro.

The Chagga have mastered a form of cultivation on the sides of their mountain which is virtually self-sustaining. They have left some of the great forest trees to provide shade for their food crops from the fierce tropical sun. Their leaves also provide some fodder for the farm animals,

174

and mulch to cover the soil. In addition, the trees supply fruits, nuts, medicines and, when they are felled, timber and firewood. Their roots prevent the soil from being washed down the mountainside. And, importantly, the big trees are only cut down if young ones have been planted to replace them.

Fifteen different kinds of banana trees are grown, for food, fodder and for brewing beer. Their roots, too, stabilise the soil. Other fruit trees – papaya and guava – are also grown for home consumption and for sale in local markets. The coffee bushes provide a useful income for the farmers and much-needed foreign exchange for the country.

Below these, taros, yams, sweet potatoes, potatoes, beans, onions, tomatoes and aubergines are grown. The multi-storey system of cultivation evolved by the Chagga is a remarkable example of applied ecology. It is based on a clear understanding of the active inter-relationship between many different components – light, shade, soil, water, trees, bushes, vegetables, animals – which has been accumulated over centuries. The crops grown in this system have been assembled from many places, including South America, India, South-east Asia and, of course, Africa.

The Chagga use virtually no mineral fertilisers in their home gardens. They keep cows, goats, pigs and chickens in small numbers which live in sheds. These are fed with vegetable matter grown in the *shamba* and grass from land higher up on the mountain. Their manure is applied to the soil of the *shamba*.

Each home garden has a network of drainage and irrigation furrows which is fed with water from the mountain. The water supply is managed co-operatively by all the smallholders who benefit from it. The complexity of the Chagga irrigation system was the object of great admiration by European travellers in the nineteenth century and even modern irrigation engineers marvel at its sophistication. Every smallholder gets his share of water from the main channels by opening an access point through which it can be let into the furrows that feed his home garden. The use of water is rotated between smallholders according to need, ensuring that everybody gets enough.

The average size of the home gardens is 1.7 acres (0.68 hectares) but, since population growth is around 3 per cent a year and every family has about ten members, ways have to be found to increase production from the available land. The Chagga are exposed to conflicting advice: some *experts* have advised them to use chemical methods for increasing output from their land, that is, fertilisers and pesticides. In fact, pesticides have increasingly come into use in recent years to combat insect pests and a fungal disease affecting the coffee bushes. But now it has been found that the irrigation ditches which used to support an abundance of fish no longer do so. The fish have been poisoned by the pesticides and have disappeared. The Chagga have lost an important source of free protein.

Agro-forestry researchers advising the Chagga have suggested that they might reduce the use of insecticides in their *shambas* by planting insect-repellent plants among their crops. This is, in fact, a traditional practice which could be enhanced by the introduction of new pest-repellent plants. In addition, the Chagga could make use of 'integrated pest management' techniques which are now proving effective in field-trials throughout the world.

In order to enhance soil fertility, agro-forestry advisers are suggesting that the Chagga could grow more leguminous plants in their gardens, notably *leucaena* trees which enrich the soil around their roots with nitrogen and whose leaves can be used as animal fodder.

The great sophistication of the Chagga cultivation system is now widely acknowledged. The integration of forest trees, food-bearing trees, bushes and vegetable crops is a remarkable achievement of applied ecology. In this age of soil erosion and galloping desertification, people throughout the tropics can learn a great deal from Chagga farming practices. Not everywhere are conditions as favourable as on the slopes of Kilimanjaro, but agro-forestry methods can be adapted to a wide variety of climatic and soil conditions.

Traditional agro-forestry systems still survive in countries such as India, Thailand, Indonesia and Brazil and they are now being subjected to close scientific scrutiny. With populations increasing rapidly throughout the Third World, sustainable systems of cultivation which do not rely on large quantities of external inputs such as mineral fertilisers and chemical pesticides must be developed and rediscovered.

Until a few years ago, the green revolution with its high-yielding varieties of wheat, rice and maize was considered to be *the* solution to the Third World's food problems. So confident were the world's leaders of the potential of these hybrid crops that they were quite prepared to mortgage their countries' earnings potential to acquire the 'green revolution package'. This usually included the construction of huge dams for large-scale irrigation schemes and for power generation, the construction of fertiliser and pesticide factories, the import of tractors, combine harvesters and farm machinery and the purchase of the new, high-yielding 'miracle' seeds.

Wherever dams were built, fertile valleys were drowned: some still forested, others were populated by nomadic pastoralists or villagers who lived by farming and fishing. These people were always promised a better life based on farming newly-irrigated lands or fishing in the great new lake. So convinced were political leaders of the marvellous potential of large dams and systems of irrigation that they were usually quite unwilling to listen to any objections voiced by people affected by such projects. Abel Alier, Sudan's Southern Regional President, discussing plans in the 1960s for a large-scale irrigation project in an area inhabited by the pastoralist Dinka, stated, 'If we have to drive our people to paradise with sticks, we will do so for their good and for the good of those who come after us.'

This sort of sentiment is typical of the high-handed attitudes of ambitious technocrats who care little about traditional life-styles which they consider outdated and no longer viable.

Edward Goldsmith and Nicholas Hildyard, in their book *The Social and Environmental Effects of Large Dams*, have amassed a large amount of evidence to show that dam projects, which are such an essential part of green revolution technology, are not the solution to the world's hunger problem. The authors argue that wherever large dams have been constructed more problems have been created than solved. Hundreds of thousands of people have been displaced from their ancestral lands. Often they are forced to resettle on hillsides in the catchment area of the reservoirs, deforesting steep slopes to create new farmland. Premature silting up has become a common problem, shortening the expected lifespan of reservoirs around the world.

Fish yields from man-made lakes are usually much lower than predicted. The lakes provide excellent breeding grounds for malarial mosquitoes and for snails that transmit bilharzia. Disease has been on the increase wherever reservoirs and irrigation schemes have been constructed. Communities affected by dam projects are invariably under great strain: 'Malnutrition and disease are rife, jobs almost impossible to find. It is a world far removed from the "paradise" offered to them by the authorities. Unfortunately, it is a world in which most of them will spend the rest of their lives.' (Edward Goldsmith and Nicholas Hildyard, *The Social and Environmental Effects of Large Dams*.)

Perhaps the most worrying problem of irrigated agriculture is salinisation of the soil. This was so in Mesopotamia thousands of years ago and it is still so today. The Food and Agricultural Organisation estimates that at least 50 per cent of the world's irrigated land now suffers from salinisation. Others put the figure even higher. Goldsmith and Hildyard claim that as much irrigated land is now being taken out of production due to waterlogging and salinisation as is brought into production by new irrigation schemes. In Pakistan, India, Iraq and also the USA, huge land areas are now affected. The problem could be solved if drainage ditches were built to let water out of irrigated fields instead of allowing it to evaporate. But this is usually too expensive to make it cost-effective, given the short-term thinking that is at the root of economic planning.

The purpose of the plant breeding programme which led to the green revolution was quite clear: varieties of staple crops were to be developed which were capable of giving higher yields than traditional strains. This seems to be a laudable aim, considering the need to increase food production for an ever-expanding world population. But there is a catch: the new high-yielding varieties require a great deal of water – hence the need to build dams and irrigation channels – and they rely on large dosages of mineral fertilisers and pesticides. Because the seed heads are heavier, the straw has to be short; herbicides have to be applied to eliminate competition by weeds which grow taller than the new cereal strains.

The green revolution brought higher yields to millions of farmers but it also created a new dependence on bought-in inputs. Those farmers who were able to afford the package increased their profits and this enabled them to mechanise their farming and to reduce their labour requirements. It is generally recognised that the green revolution has contributed considerably to rural unemployment and that it has accelerated the drift to the cities by rural people throughout the Third World.

There is no doubt that the green revolution enabled the larger farmers to expand their operations at the expense of smallholders and landless labourers. Susan George, in her book *How the Other Half Dies*, quotes a former minister of agriculture in India:

> Mr Jagjivan Ram . . . has been reported as saying that in his country the beneficiaries of the Revolution are not the peasants 'who live miserably on a few rupees a month' but the small, privileged stratum of large landholders. While 22 per cent of rural families own no land at all, and 47 per cent own less than an acre, 3 or 4 per cent of the large proprietors with political power and influence are in a position to appropriate for themselves all the resources in inputs, technical assistance and credit put at the disposal of farmers by government agencies.

The substitution of high-yielding varieties for traditional local strains has had a dramatic impact on the genetic variety of the major food crops. Wherever the green revolution has taken a foothold, traditional strains are on the retreat. But plant breeders need genetic variety to be able to breed new hybrid, high-yielding strains. Often traditional varieties have a greater resistance against plant diseases and are indispensable for plant breeders. But as the new varieties take over, the genetic base of the main staple crops is further and further eroded.

Global 2000, the world study commissioned by President Jimmy Carter, expresses concern about this state of affairs:

> If present trends continue, increasing numbers of people will be dependent on the genetic strains of perhaps only two dozen plant and animal species. These strains will be highly inbred, and the plant strains may have reduced pest and disease resistance and may be planted in large, contiguous monocultures. Plant and animal epidemics will occur as they have in the past, except that in the future the number of human lives at risk may not be in the millions (as was the case in the Irish potato famine), but in tens or even hundreds of millions.

Pesticide resistance of insects, weeds and fungi is becoming an increasing problem worldwide. Over 400 insects and mites were resistant against one or more insecticides by 1980. Weeds and fungi, too, are increasingly able to withstand pesticide applications, making ever more frequent and varied treatments a costly necessity. The USA, as usual, is showing the way. Cotton crops in parts of the USA now require thirty to fifty applications in a single growing season. In some districts, farmers have had to give up growing cotton because they were simply no longer able to defeat the pests in a cost-effective way. Much the same situation occurred in cotton-growing areas in Mexico, Nicaragua and Sudan, where the increasing costs of pest control made the production of cotton less and less economic.

Resistance problems with insect pests, notably the 'brown planthopper', affecting rice are assuming major proportions in South and South-east Asia. This insect was a minor pest until high-yielding rice varieties were introduced as part of the green revolution package. Since these varieties take much less time to mature, two or three crops of rice a year are now grown. The planthopper thus has a continuous food supply; it has become a very persistent pest, requiring ever more pesticide applications to keep it under control. These in turn have reduced natural predators, thus further aggravating the problem.

Research by scientists of the Bundesanstalt für Naturschutz in Bonn, West Germany, has shown how broad-spectrum insecticides affect insect populations. They sprayed a field of beans which was mildly affected by aphids. A day later, the scientists searched the soil for dead insects. Of these insects, 94 per cent were useful insects such as bees, ladybirds, beetles and spiders. None of the beneficial insects survived the treatment, whereas some of the aphids stayed alive and soon started to multiply again. Before the treatment, there were eighteen aphids for every ladybird. (Ladybirds each eat about thirty aphids a day.) Six weeks later, the scientists found ninety-one aphids for every ladybird. This story illustrates clearly how insecticides reduce the natural predators of insect pests and ultimately help the pests to multiply. And that is the beginning of the 'pesticide treadmill' which has become a reality for millions of farmers around the world.

First, the insecticides kill off the natural enemies of insect pests which are thus able to multiply rapidly. More pesticide treatments reduce them further but tend to enhance the survival chances of those insect pests that have inbuilt defence mechanisms against the poison. The insecticide treatment thus contributes directly to the selection of resistant strains of insects, making it more difficult for the farmer to defend his crops against the insect pests.

Consequently, ever more applications of an ever greater variety of insecticides become necessary, increasing the chances of farmers endangering their health through exposure to the poisons. This is a great problem in Third World countries where pesticides are usually applied with knapsack-sprayers and where people often cannot afford suitable protective clothing.

David Bull, now director of the Environment Liaison Centre in Nairobi, has made a study of the pesticide treadmill in his book *A Growing Problem*. Citing examples from many Third World countries, he has shown that the spiralling use of pesticides is not the solution to the food problems of the developing countries. He emphasises the addictive nature of pesticide use and the growing hazards to food producers and consumers. He shows how pesticides which have long been outlawed in Europe and the USA are still used in the Third World and find their way back to the West in imported foodstuffs.

Bull does not deny that pesticide use has increased levels of food production in the Third World but he deplores the social and environmental costs that have arisen from this. He cites the example of wet rice cultivation: traditionally, people in countries such as Indonesia, Malaysia, the Philippines and Bangladesh used to catch a lot of fish in rice paddies, which was a major source of their protein. Today, the intensive use of pesticides in rice paddies has resulted in a dramatic decline in fish population. In some places, fish, crabs and snails have disappeared altogether. In Indonesia alone, fish were harvested from 7.4 million acres (3 million hectares) of rice fields with a potential annual production of 600,000 tons. Today these have all but vanished, leaving a major gap in people's diet.

Annual pesticide use worldwide is about one pound (or 0.45 kilograms) weight for every person on Earth. The trend is still upwards, even now more and more farmers are getting on the pesticide treadmill. Pesticide poisoning of farming families and farm animals is reaching epidemic proportions in the Third World. Some 10,000 people are thought to die from it every year. Hundreds of thousands fall ill and often the true cause of their sickness is not diagnosed. The terrible explosion at the pesticide factory at Bhopal in India in 1984 killed 2500 people according to European press reports, although other sources put the figure much higher. Tens of thousands of people will suffer from lung disease for the rest of their life. Is this a signal that it is time for changing course?

Integrated pest management is now widely advocated as a sensible alternative to the pesticide treadmill. The entomologist Professor DeBach gives this reason in his book *Biological Control by Natural Enemies*,

The empirical and unilateral use of chemicals to hammer pests into submission by repeated costly blows is . . . increasingly failing to provide a solution. A rapid and drastic change is necessary in order to achieve control of pests in an ecologically and economically satisfactory manner.

This view is now echoed by pest researchers around the world. Integrated pest management uses biological control methods as the first line of defence against pests. Suitable crop rotations are adhered to in order to discourage the build-up of pests. Predatory insects are, if necessary, deliberately introduced to reduce insect pests. Insect-eating birds are encouraged by providing suitable habitats. Hormone traps and light traps are put up in the fields to attract harmful insects. Pesticides are used only as a last resort.

The aim of integrated pest management is not to eliminate pests completely since this has proved virtually impossible anyway, even using the most sophisticated pesticides, to keep pests at manageable levels. Field-trials on a variety of crops have shown that this aim can be accomplished. Farmers throughout the world will be encouraged to use their traditional knowledge as the basis from which to adopt more recent techniques of integrated pest management. Practical tests in China, Sri Lanka, India and Nicaragua are yielding encouraging results.

The majority of the people of the Third World have only begging left as an option if they do not have land on which to grow their own food. Virtual destitution is the daily reality for hundreds of millions of people. In 1974, Henry Kissinger told the World Food Conference in Rome which was attended by government representatives from 150 countries, 'Within a decade, no child will go to bed hungry, no family will fear for its next day's bread, no human being's future and capacities will be stunted by malnutrition.'

Ten years later, the world's hungry people have increased from 400 million to an estimated 700 million. In some Third World countries, notably India and China, the food situation has improved, partly as a result of green revolution advances. But in large parts of Africa and South America the situation has got far worse, Ethiopia being the most widely publicised example. This is how Mohamed Amin, the Kenyan journalist who made the BBC film that first drew Western attention to the famine in Ethiopia, described what he saw at the end of 1984:

There was this tremendous mass of people, groaning and weeping, scattered across the ground in the dawn mist . . . as if a hundred jumbo jets had crashed and spilled out the bodies of their passengers among the wreckage, the dead and the living mixed together so you couldn't tell one from the other.

Food aid has come too late for many, we shall probably never know how many. In north Africa, the future for so many will not only be determined by the outcome of wars, but also by sufficiency of rainfall and by the capacity of depleted soils to yield harvests. The hills of Ethiopia today are treeless where only forty years ago there were large expanses of forest. The soil is no longer protected from rainstorms or desert winds. The sun has become merciless as the shade of trees no longer protects soil, people or beasts.

If only it were possible to transplant the tree groves of the Chagga from the slopes of Kilimanjaro to the highlands of Ethiopia. Indeed, similar systems of cultivation were once practised there; agro-forestry, the great new hope of Western aid organisations, is an age-old tradition in tropical countries. The oases in the Sahara, those that have not yet dried up or been overwhelmed by the sand, are witness to this: they are shaded by the multilayered canopy of palms, fruit trees and food crops.

Right across the Sahel, drilling rigs are probing for water. The fossil water that is found and pumped up can be as old as 30,000 years, dating back to an age before the invention of agriculture, agro-forestry or cattle herding. Nobody likes to imagine what will happen to the remaining trees and vegetation on the edge of the desert if pumping results in a depletion of the groundwater.

Climate researchers are cautious about future prospects for rainfall. Dr Michael Dennett, a climatologist at Reading University, has recently completed a study of rainfall changes in West Africa over the last forty years. These are some of his conclusions:

The Sahel's rainfall from 1974 to 1983 was about 5 per cent less than in the 1931 to 1960 period. We cannot, however, be certain of a trend towards a long-term decline. . . . But one finding was statistically significant. We analysed the Sahel on a month by month basis and found that, undoubtedly, during the last twenty years, rainfall in the main rainy month of August was around 10 per cent lower compared with rainfall in the Augusts of the period 1931 to 1960 . . . It is starting later and ending sooner.

One thing is certain: the pressure on the trees still remaining in the Sahel is great, people are desperate for firewood and for animal fodder. On its southern edge, in Nigeria and the Ivory Coast, deforestation for firewood, for clearing land to be used for farming and for tropical hardwood exports is now reaching 10 per cent a year. The impact this has on reducing atmospheric moisture and raising soil temperatures can easily be imagined. The pace of deforestation in neighbouring countries such as Liberia, Ghana and Sierra Leone is not much slower. Is this contributing to changes in the rainfall pattern?

For untold millions in the Third World, hunger results not only from lack of adequate crop protection, from lack of land, fertile soil and water, but also from lack of trees. Removal of trees is a precondition for creating farmland, for tree cover is – or was – the 'natural condition' of much of the land surface of the Earth. But removal of too many trees is a danger to agriculture because it encourages soil erosion by water and wind and it can contribute to climatic change. We saw earlier how this process affected the

Mediterranean countries in the past. Today, large parts of Africa and, increasingly, Middle and South America are suffering under rapid deforestation. In Africa, the rain forests of Zaïre and Cameroon still cover huge areas. But they are shrinking at a rate of an estimated 1. 3 million hectares a year, a process which will be accelerated once new roads which are being built for timber extraction are completed.

Africa, and much of the rest of the Third World, is caught in a traumatic situation: forests are being sacrificed to create more farmland. Valuable hardwood trees are being felled to bring in foreign exchange. Some of the best land is under cash crops for export to Europe and the USA. The ever expanding cities require cheap food: fertility is taken out of the soil, never to be returned. Development aid is backed up by expert advice which has encouraged European style 'open land' farming not suitable for tropical conditions. Sheet and gully erosion are rife not just in Kenya, as John Seymour has shown, but in much of Africa.

Population pressure under these conditions results in a breakdown of ecosystems. In Mauritania, another country on the southern fringe of the Sahara, more than 80 per cent of the grazing land used by nomadic pastoralists has turned into desert. Towns are now ringed by tent cities inhabited by people who have lost their livestock, seeking food. Few Westerners know anything about Mauritania, the television crews go elsewhere. But Mauritanians know about hunger as well as do Ethiopians.

On the outskirts of Nairobi in Kenya, we saw where peasants without fertile land and pastoralists who have lost all their stock end up – if they survive. We drove straight into a slum in our shining cars. We stopped and people congregated all around us; there were some smiles and many clenched fists. Ali Twaha, our Kikuyu guide, was verbally assaulted by the young men: how dare he bring in whities with their film cameras and tape recorders? Wasn't it people like us who had brought about all this misery? On whose side was he anyway?

Ali stood his ground while we filmed the camp, which was on the side of a valley, shacks made of rusty sheets of corrugated iron everywhere. Many people were in rags, a few still wore colourful tribal dress. There were children all around, learning how to be angry and resentful at an early age.

There are few jobs for destitute peasants in Nairobi, one of the better-off cities in Africa. Former pastoralists like Boran or Masai are preferred as guards and night-watchmen in office blocks occupied by international organisations and government departments. Otherwise, employment is scarce.

In these buildings and their counterparts in London, Washington, Paris and Bonn, the destination of the people of the Third World is being decided. Grand visions of the future have evaporated in the face of reality.

Political leaders are no longer talking about creating a new paradise. Today, the goal of well-meaning politicians is simply to help the people to survive in dignity. Julius Nyerere of Tanzania stated, when asked what was the greatest accomplishment during his time in office, 'That we have survived.'

There is much bitterness in the Third World about the low prices paid by the rich countries for raw materials and for cash crops. A large percentage of export earnings simply pays for servicing foreign debts. Many countries do not have the funds to develop industries which would make them self-sufficient even in essential machines or consumer products.

The mirage of a world dotted with great dams, fields covered in pest-free crops, machines instead of people farming the land has remained beyond our grasp. In Africa, a viable system of cultivation on millions of small peasant farms is the most important goal to work towards. Just to keep the soil from blowing away must be the most immediate task. Are we at last prepared to help the people help themselves?

The Plight of the Animals
—————— John Seymour ——————

We went to Austria principally because we wanted to see for ourselves an agriculture and an animal husbandry that is still fairly traditional, wholesome and humane. Fine brown, or brown-and-cream-coloured cows graze in that country on the splendid upland pastures to produce magnificent milk (far richer than the watery liquid that is produced by the black-and-white animals in countries further north) and most of this milk is made into some of the best cheese in the world.

We drove to the big village of Reith, in the Alpbach Valley of the Austrian Tyrol. The valley was fairly level, some of it was ploughed, growing mostly maize which was being cut with forage harvesters for silage, some of it was lush permanent pasture. Most of the houses in the village were farmhouses and for my money they were the finest farmhouses in the world. Built entirely of wood, roofed with wooden shingles (we watched a man splitting these with a *fro*, a small hand tool, from larch logs), these houses are a farm all under one roof. The south end of the building is the dwelling portion, and consists of a three- or even four-storey house with balconies on the south end, and the balconies are almost always heavily loaded with beautiful flowers. The great roof covers this portion and also covers the farm part of the building, which occupies probably four-fifths of it: from the dwelling part to the north gable end. In the loft goes the hay – stuffed through a door high up in the north gable end. Below the hay-loft are the cows, comfortably bedded down for the winter and living on the hay which is flung down to them from above. On the sides of the building outside, sheltered by the overhang of the one great roof, is firewood, tons and tons of it, all sawn, split and neatly stacked. You feel that the whole great building is perfectly designed to shelter people in warmth and comfort, and cows, and fodder, and firewood, all through the cold and snowy alpine winter. All under that one magnificent shingle roof. If the place were beleagured by snow for months on end the work of the farm could go on, milking, cheese-making, feeding cows, rearing calves, without anybody putting his nose out of doors at all.

But we went there at the end of September, and in September the cattle were far away, most of them high up on the alpine pastures. To make use of the herb-rich mountain pastures, the cattle are herded up, in the spring and summer, where possible right up to the snow-line.

Traditionally, on 1 October the cows come down to Reich. We made friends with a young farmer, who lived with his family in a typical farmhouse in which he kept guests from the city as well as cows. He led our convoy of cars far up the valley, and then by a small side-road into the mountains. Higher and higher we went until we came to a superb green valley, or hanging plain, surrounded by steep mountainsides. Here we found a mountain farmstead: a wooden chalet, in which the cowmen lived during the summer; a big cowshed in which the cows were tied for milking; and a dairy in which cheese could be made. A cableway led up the steep mountainside to the heights above and we were told this was for sending down the milk every day.

We were met by a fine old man with a huge bushy beard, dressed in traditional Tyrolean dress: a brimmed hat with a feather in the side, a collarless green jacket and breeches. He was one of nature's bachelors. I do not think he was capable of forcing his face out of the configuration of a merry smile. He had lived up there with the cows the summer through, herding them and milking them, helped, for the most part, by his young boss, the farmer.

Leaning against the wall of the chalet were strange objects. They were like big fans, or plumes, made of frameworks of pine bows covered with artificial flowers of gaudy colours. We saw the men (and one small boy) catching the cows, one after another, as they came out of the cowshed and tying one of these frames on to the horns and heads of each one. The cow was then released and she ran, or galloped, down into the valley to join her sisters, who were already decorated with these objects, and they were all behaving in a most extraordinary way. I swear they were like rather vain young ladies trying on new ball dresses! They admired each other's head-dresses, they tossed their heads, they pranced and galloped and cavorted. Each cow had a bell around her neck and the noise that all these bells made (carefully chosen as they had been for their harmonising tones) was delightful.

Now lest it be deduced from all this that cows are *vain* animals let me explain that all the cows – except for a few young heifers – knew perfectly well what it all *meant*. They knew they were being dressed up to go down to the valley below at Reith, where they would first have a couple of months grazing on lush valley pasture, fed perhaps as a supplement on fresh-cut maize, and then spend the winter, when snow was about, in a warm, well-strawed house being fed on silage and the finest hay. They felt – and behaved – like children at the end of a term at a boarding school.

When they were all decorated and assembled, the fine old man with the beard started proudly leading the way down towards the little road. The ladies followed him. Their leader, always, was a magnificent old girl with the biggest bell tied to her neck – it was a great iron bell as big as a large pumpkin. The farmer told me that once, because she – the matriarch of the herd – had recently calved and he thought her a bit too weak to lead the herd, he had transferred her bell to a younger cow. The former queen had promptly rammed the upstart in the side, nearly knocking her over, and behaved in such an intolerable way that the men had quickly transferred her bell back to her – and with it the symbol of the leadership of the herd.

Along they all trotted, the old man leading, the little boy running where he was wanted, the young farmer stalking along behind. There was no need to *drive* the cows: it was not that sort of relationship. You always notice that when men and grazing animals have much to do with each other, the man – shepherd or cowman – walks before the herd or flock and the animals follow him. Our old cowman friend, by the way, conspicuously carried a bottle of schnapps, from which he frequently drank.

Well, it was twenty miles back to the farm in the main valley, and we eventually got there, in our convoy of vehicles, and waited on the road for the caravan to appear. The cows, though tired, scented home and picked up strength amazingly. Then we saw them turned into a broad field of good grass and clover. They galloped about for the joy of it but soon settled down and had their noses buried in the rich pasture.

Most of the population of the village, including the holiday-makers and tourists, had turned out to welcome the cows. The next day there was a grand celebration in the church, a parade of the local defence force, the Scouts and the Guides, the blessing of the fields and other observances and jollifications. But there was no doubt whatever who the stars of the whole performance were: the brown-and-cream ladies with the lovely head-dresses.

And why not? Apart from the tourists they were the source of prosperity for the whole region. They were capable of doing what no man has ever been able to do: turn grass into good cheese that people can eat. And they enjoyed doing it too, as the men obviously enjoyed working with them.

We then went to see a beef lot at La Torre run by the Societa Co-operative Agricola Zootechnica, near Isola Della Scala, on the great, flat, fertile plain of northern Italy. We went to see where all the farm animals have gone for, by and large in this area, they are not on the farms.

Here we find 6000 bulls. A matador's wildest dream, one might think, but these bulls would have provided him with no sport because they showed every sign of having been heavily sedated. They were quite silent except, very occasionally, one screamed.

On thirty-seven acres (fifteen hectares) of flat land there were twelve sheds, and each shed contained between 500 and 600 cattle. Except for one shed, which is for new arrivals to this boys' finishing school, there are no outside yards for the bulls to walk about in. It is almost true to say that they cannot walk at all because they are pretty crowded. But they are not there to walk about – they are there to put weight on, to make profits for the co-operative. Over the heads of the cattle, very near to them, is a mesh of electrified wires. This is to stop them from mounting each other. All cattle like to do this occasionally, either from sexual frustration or from sheer high spirits. The animals at La Torre quickly learn not to do it. They are not allowed high spirits. Sexual frustration they must have, for they have not been castrated.

To look after them are eleven ordinary men and five vets. The latter are necessary because animals living in such high stress conditions have to be kept under constant veterinary supervision.

I walked along the gangways of several of the sheds and found it a very disquieting experience – in fact I doubt if I shall ever forget it.

My reactions were strengthened by the fact that I had cattle of my own for about twenty-five years and therefore I cannot consider them just as a *means* to human profit or nutrition, they will always seem to me to be also *ends* in themselves. It is no good telling me they have no 'feelings'. I know they have feelings: of affection, of maternal and filial love, of comradeship to other cattle, of jealousy, and – yes – of sheer *fun*. These 6000 bulls just looked at me with uncomprehending misery in their eyes – why should this fate have happened to them – why should they, in their short lives, be cut away from everything that means life experience to an ox: green grass, running, playing, feeling the sun on one's back, hating the rain, being pestered by flies? For life experience cannot all be pleasure. But my feeling as I walked along those endless gangways was one of intolerable shame. I felt ashamed to be a human.

When the first man or woman adopted the first orphan goatling or lamb, and brought it up to maturity, to keep as a pet, or to milk, or to eat, or whatever, the human concerned was interfering with the *natural* life-style of the animal. Since then, all animal husbandry has been 'artificial' to a certain extent. But it is all a matter of *degree*. If you allow a bull calf to suckle from its mother for three months, then run out on grass fields with other cattle for a year or two, then go into a fairly spacious shed, with twenty or thirty companions, well-bedded on dry, clean straw for a winter, then be driven to a local slaughterhouse and humanely killed, you have arranged for the animal to lead a not completely natural life, but a not completely unnatural one either. It was perfectly happy in its bovine way during its lifetime and, if its ancestors had not been domesticated and it had been born in the wild, it would probably have been killed by a pack of wild dogs or a lion. And an animal so treated could live its life out without one grain of either sedative or antibiotics.

Later, Herbert Girardet interviewed Professor Englehard Böhnke, of

Kassel University in Germany, who is a veterinary scientist, and who like many other knowledgeable people is very disturbed about animal factory farming methods. Here is a little of what Professor Böhnke said, translated into English:

We can see that modern intensive farming systems, where stall design is concerned, are often constructed with little regard for the welfare of the animals. This means that in many cases the animals cannot enjoy the behaviour pattern that is characteristic of their species. And this, we know, is an additional stress factor that weakens the power of the bodily organisms to defend themselves . . . Their being frustrated in their desire to follow the behaviour pattern of their species, and other stress factors, reduce immunity to the point when diseases are again contracted, which in turn have to be fought again with drugs . . .

He went on to list the diseases which intensively-kept cattle are heirs to and said:

It is difficult to see how we are going to get out of this problem. It has got so much worse . . . that experts now talk about intensive farming systems as if they were a kind of hospital system. That is to say a situation that is comparable with the situation in hospitals where there are large colonies of harmful bacteria in a restricted space, which can only be combated with difficulty . . . medicaments, that is, in the form of antibiotics, cortisone preparations, tranquillisers, and so on. This growth in the use of drugs has not only disturbed the public – it has also led to ecological problems on a far greater scale. We are having problems today over the use of antibiotics because many disease-producing bacteria have become resistant, in other words they no longer respond to these substances . . .

Certainly, intensively-kept livestock of all sorts are suffering from diseases that did not even exist when I was at college. So far, the scientists have been able to keep just one jump ahead of the pathogens. But bacteria, viruses and even insects seem to have an extraordinary capacity for adapting to ever stronger and stronger antibiotics. My own experience, from twenty-five years of farming my own land and many years of working with other people's animals is that, if you keep animals in conditions reasonably close to the conditions in which their ancestors were evolved to live, then they hardly ever get ill.

The film crew for the television series managed, after great difficulty, to film in a hen battery house. Owing to intense bad publicity more and more owners of these places will not admit the public or the press but, after making various promises, our producer managed to gain entry into one. I did not accompany him. I had seen enough of battery houses. I used to accompany an old friend of mine in Suffolk, who was a poultry dealer, on his rounds buying the 'scrapped' hens from battery cages. The hens had all been in the cages for less than a year, for they are scrapped at their first moult. In any case they would not last any longer – most of them had growths on their chests from leaning on the wires, and their livers and other organs were diseased. We used to stuff them into crates by the thousand and take them to a packing station north of Ipswich. The remains of these diseased birds used to go to make canned soup. I kept a few of the hens once or twice to see if they could adapt to living on free range. It took several days for them to learn to *walk* for they had never done it in their lives. Scratching in the ground, too, was something they had laboriously to learn – they had never been on the ground. It was several days before they were able to flap their wings. They were pathetic creatures – they reminded me of men who had been put to the rack.

Twenty years ago the British government set up a committee to inquire, among other things, into the humanity of keeping hens in battery cages, and of course, as everybody knew it would, the committee reported in favour of battery cages. In 1965 the Brambell Report was published. It stated:

The behaviour pattern of the modern bird remains essentially that of its wild ancestors. It is still a bird of gregarious habit, that establishes and maintains a high degree of social order within the group, the members of which can recognise each other and communicate vocally. Although the birds spend most of their time on the ground scratching and searching for food, they can and do fly. Even the modern bird can fly and will do so if the situation requires it, contrary to some assertions we have heard. Maternal care for, and instruction of, the young is highly developed.

After this the Brambell Report said that battery cages were all right in principle but they should be big enough so that a bird could stretch out *one wing*.

Here now is a quotation from a pamphlet, put out by Compassion in World Farming, called 'Battery Egg Production'. It is correct in all details:

Four-bird cages are the norm, measuring 16 ins by 18 ins and allowing only 72 sq. ins of floor space per bird, though five-bird cages of 18 ins by 20 ins are also common. The average wing-span of a chicken is 21 ins, but under current stocking densities they are allowed only 4 ins of cage space per bird. They cannot fulfil any of their innate behavioural instincts. They are unable to perch, fly, scratch about on the ground, spread their wings or make a dustbath. The only conceivable exercise is one step to the other side of the cage and this is only possible if the other three occupants don't get in the way . . . The light intensity is maintained for up to eighteen hours a day as this stimulates egg production. Unfortunately it also stimulates feather pecking and cannibalism because of boredom and in these circumstances the only remedy is to debeak the hens and reduce the light intensity.

It is perfectly legal for operators, in England at least, to advertise eggs produced in these conditions as 'Farm Fresh Eggs'.

I will turn now to pigs. The story of pig keeping can very well stand for the stories of the other domestic animals. It is a story of a problem being solved by a technological fix, leading to another problem, leading to another fix, et cetera.

For eight years I farmed a five-acre smallholding on very light land in Suffolk and all during that time I kept pigs. I kept them as near to the conditions for which the wild pig was evolved as I could. I kept six sows and a boar and, except for two or three pigs which I fattened myself to kill for bacon every year, I sold their progeny at from eight to twelve weeks old as 'weaners'. They nearly all went to one biggish farmer, who fattened pigs near Debenham, and he always gave me a pound a head over the market price for them because he said they were so healthy and never suffered respiratory diseases, unlike the other weaners he bought.

I kept all my pigs out of doors and I was well aware that this might give rise to certain problems. If I let them loose they would ruin my neighbour's wheat and dig up his sugar-beet. If I contained them too closely on one piece of land they would suffer from intestinal worm infestation for they would be rooting continually on infected soil.

So I confined them behind an electric fence, on a fairly small area of land, and moved them fairly often. They were never suffered to remain on one piece of land for more than three months and did not return to that piece of land until it had been ploughed up and had several crops taken off it. My six sows had two litters a year, or nearly, would produce twelve piglets apiece every time with monotonous regularity, rarely killed a piglet by laying on it (I remember this once happening), never ate their piglets, and quite monotonously reared every one. They lived and farrowed in the roughest of rough movable pens, and were provided with tons of dry straw.

There were three disadvantages to this system (which was not entirely 'natural' – electric fences are rare in nature), but not too unnatural either. One was that it took a lot of labour. My labour was free so I did not count it. Another was that my sows ate a little more food being outside than they would have done inside. This I did not mind either because I grew most of the food. The third was that, because I did not practise early weaning, my sows had slightly fewer piglets per year than if I had weaned them, as my commercial neighbours did, at, say, six weeks old. I did not do this because I thought it was cruel: a consideration that would certainly have no weight with the majority of factory farmers. Incidentally, one advantage rose from this practice of late weaning: my sows went on producing well, and keeping their health, for far longer than sows in intensive systems.

Now then, the first technological fix in modern pig keeping was a very simple one: to confine sows on one small piece of land next to the farm buildings. This saved labour. It caused a problem though – the sows suffered badly from worms. As we have seen, confining pigs on one small piece of land causes infection from worms. In intensive modern pig-farms, where land is at a premium, the practitioners have come up with an answer, or fix:

Fix – keep the pigs on concrete so that they cannot eat earth and reinfest themselves with worms.

Problem – pigs have a very powerful behavioural need to *root*. Because this was thwarted there was stress, causing disease.

Fix – medication.

Problem – piglets, out of doors, eat earth – kept on concrete they get iron deficiency.

Fix – inject all young piglets with iron.

Problem – piglets, denied rooting and running about, bite one another's tails and even eat one another.

Fix – cut off their tails and break their teeth.

Problem – sows, the chain of instinct of the farrowing process disrupted, started engaging in unpiggish behaviour, such as allowing their piglets to be smothered in the straw of the farrowing pen.

Fix – deny them straw.

Problem – denied straw (a sow will spend days before farrowing running about with bunches of straw in her mouth – starting to build a nest, now here, now there, until she finally, elaborately builds one. Having gone through this ritual she will not allow her babies to be smothered in straw) the mother will engage in even more unpiggish behaviour and lay on her pigs or – very commonly too – eat them.

Fix – make her farrow in a crate. In this she can stand up – lie down – but not turn round. She is confined like this until her piglets are weaned. Her chain of instincts is now completely shattered but this does not matter because the piglets are lured away from her – except when she is laying down and they are suckling – by an infra-red lamp. (We must not indulge in anthropomorphism – pigs are not humans, but anyone knowing what lively, active and intelligent animals pigs are must realise the revolting cruelty of keeping a sow so completely confined night and day for a couple of months.)

Problem – the piglets, thus unnaturally kept from their mothers, become prone to virus pneumonia and other ailments.

Fix – fill them up with antibiotics and other medicaments.

Problem – the piglets, kept thus unnaturally, contract bacilli and viruses from their mothers to which, because of their lowered resistance, they are unable to cope.

Fix – the system of weaning at birth being developed at the Rowett Pig Research Institute at Aberdeen. Cover the sow's head with plastic while she is giving birth and never let her see the pigs or the pigs see her. This fix is described in a recent edition of the magazine *Pig Farming*:

In the new unit, an automated system is being developed to remove piglets – identified by an electric grid – on a slow-moving rubber conveyor belt. This takes the piglets through a trapdoor in the wall, and into the adjoining rearing room. Low-intensity, ultraviolet light kills any bacteria remaining in the trap.

Elaborate precautions are taken to prevent dams passing on 'bugs' to their offspring – down to polythene sheeting over the sows to ensure that the newly-born piglets do not even breathe the same air as their mothers.

The piglets, having been thus conveyor-belted away from their mothers' wombs, are then each put in a plastic box just big enough to hold them, fed there on evaporated milk through nipple-drinkers for two weeks, then removed to battery cages where they are kept, isolated from their kind, until they are considered old enough to be taken to the fattening pens. I was taken over a fattening unit once on the Isle of Anglesey. Here, 600 fattening pigs were crowded into cells in total darkness – except for three twenty-minute periods every day during which they were fed. The heat and humidity were kept intense, so that no food was used up in keeping the pigs warm. They were kept in total darkness so they could not see to fight each other.

Problem – in the Rowett system it is found difficult to make the mothers take the boar again after their piglets have been thus whisked away from them and they have not had the experience of mothering them.

Fix – kill the sows (still known as gilts – that is, they have had only one litter) immediately and turn them into bacon.

But there is another fix even more unsavoury than the Rowett system. That is the creation of a 'minimal disease herd'. The mothers are slaughtered just before they would have farrowed naturally and the piglets cut out of their dead wombs. The piglets are then supposed to have no disease organisms and therefore can form the basis of a disease-free herd. I knew a man in Suffolk who achieved a herd of 400 'minimal disease' sows. Swine fever broke out in his herd (which had been created at enormous expense: the caesarians have to be performed by vets), and the whole lot had to be slaughtered.

This cycle of problem and fix has only been going on for a few decades. One wonders what the state of the art would be, if it were to continue, by the year 3000? Is it frivolous to suggest that the whole industry will by then have become so complicated (and so palpably cruel) that some genius will try, well, keeping sows out of doors again, giving them plenty of straw to make their nests with, allowing them a reasonable time to suckle and mother their children, to recover after their pregnancy, and cut out all the antibiotics, oral and subcutaneous vaccinations, iron injections, tail amputations, tooth-breaking, solitary confinement, anabolic steroids and all the rest of it? Maybe pig farmers would have to do some physical work in the open air. That wouldn't hurt them. Maybe they would make a little less profit. But what profiteth it a man if he gain the whole world and lose his own soul . . . ? I am certain the bacon would taste a lot nicer. Another consideration is that the manure would go back to where it belonged – on the land. When the world's phosphate deposits have all been squandered and nitrogen costs are enormous because the oil has run out maybe this factor will be of some consequence.

The subject of manure takes us back to the gigantic beef lot in northern Italy. What do they do with the manure there (and believe me, 6000 bulls produce a lot of manure)?

Well, the bulls are kept on slats and the manure falls into channels underground. Some of it is pumped into tanker lorries and taken away to be spread on the fields that produce the maize that the bulls eat. Eighteen local farmers, who together farm 1750 acres (700 hectares) of land, provide the maize and are supposed to take the manure. They take some. Raw manure is a disgusting substance and there are severe problems about putting too much of it on the land.

During our visit I wandered away, unnoticed, to the side of the enterprise where we had rather obviously not been taken. And there, stretching for hundreds of yards away from us, was an enormous lake. Beyond the artificial *bund* on the other side was another equally enormous lake, obviously constructed because the first one had not proved big enough. Both were filled to overflowing with what the agribusiness world likes to call slurry but for which I have a shorter name. The stink was awful. It was quite obvious that the eighteen farmers, who were supposed to take the stuff away, could deal with only a small fraction of it. The stuff would simply be left there, as slurry is in thousands of such beef lots all over the world, until most of the liquid from it had seeped down through the soil to pollute the underground water below, and the solid matter had rotted away.

I remembered the strong wholesome smell of the farmyard manure that I helped cart from the bullock yards back in Essex and back in another age, and I remembered the fine feel and quality of the well-manured land. But farmyard manure took *straw* to make, and *labour* as well, in quantity. Nowadays, nearly all the straw is burnt, and as for the other component – you have to look at the dole queues in the cities to find that.

The Organic Alternative
———— John Seymour ————

To bemoan the evils of inorganic agriculture: soil erosion and degradation, nitrification of food and water supplies causing at least part of the rising rates of cancer, depletion of fossil fuel supplies, dislocation of rural society, increased urban crowding and unemployment, greater dependence of the dwindling farming communities on chemical and machinery firms and money-lenders, disruption of wildlife and so on – is useless unless there is a practical alternative.

We went to the farm of Mr and Mrs C. B. Wookey, near Upavon in Wiltshire to find out if there is.

There is. Actually there are now over 500 successfully operating commercial organic farms in England and Wales alone, and far more in Germany. In all Western European countries the number of organic, or biological, growers is increasing. A German biological farmer said to me, 'All right – we are still a small army but it's an army that is growing. I had rather be a soldier in a small army that is growing than one in a large army that is diminishing!'

We drove into the yard of the large and ancient manor house in which the Wookeys live. One thing was obvious: these people were not on the breadline. Large shiny horses looked at us over the half-doors of stables, a large shiny Land Rover stood patiently waiting for its owner. Inside, the house was as beautiful and gracious as it was outside.

After coffee I climbed up into the Land Rover beside Mr Wookey, who, I noticed, took his thistle-spud with him – a long stick with a tiny spade on the end of it for rooting out thistles. We drove out of the little village and up on to the Wiltshire chalk downs.

There was a time when I would have doubted the ability of thin, light soil on chalk to grow decent wheat at all. Since the chemical revolution of course quite good crops of wheat are grown on all sorts of soils, including chalk. If you can just get a soil to hold the wheat plants *up*, and then feed them with all their requisite nutritional needs 'out of the bag' (with artificial fertilisers), they will grow and produce wheat. But on Mr Wookey's farm I saw excellent wheat grown on shallow chalk soil which had had no 'bag' fertiliser put on it for, in some cases, well over a decade.

The estate is 1650 acres (670 hectares), and there is a further 400 acres (160 hectares) of rough grazing which is rented from the War Department, on the tank firing ranges near by.

195

The estate was divided then into a biological section and a commercial section. The Wookeys, when they started to convert to biological farming in 1970, decided not to convert all their land at once. They wished, for one reason, to be able to compare the biological section with a control. In 1984 the biological section was 1245.5 acres (505 hectares), leaving 404.5 acres (165 hectares) in commercial, or orthodox, practice. By the end of 1985 the Wookeys planned to wind up the commercial section and convert all the land to chemical-free farming. The Wookeys found that they were losing too much money on the commercial section to make it possible to continue with it. The escalating cost of chemicals was the cause of this.

The method that has been used for converting the land has been this: every year since 1970, another 100 or 200 acres or so has been designated to become biological. This land has been drilled to a spring barley crop undersown with grass and clover seeds and treated with artificial fertilisers in the conventional way. Without these latter it would have been a very poor crop of spring barley. From that time on, though, this land receives *no* chemical fertilisers or poisonous sprays whatever. And after three years it is officially declared biological and the produce from it is sold under that label.

The biological rotation, which of course is all important if you want to succeed in the trick of growing good arable crops without 'the bag', has been the subject of much experiment on this farm but seems to be settling down as: four years' ley (grass and clover and herbs); two years' wheat or barley or oats – the last crop undersown to red clover; one year's red clover; two years' wheat barley or oats; and four years' ley again, which is where we came in.

The most important of the white straw crops is wheat. The yield of the latter seems to be about comparable with the yield of orthodox farms near by. We visited the Wookeys towards the end of July and I was amazed at the cleanliness and quality of the wheat. I could see absolutely nothing wrong with it at all. There were 415 acres (170 hectares) of it – modern breeds with names such as Maris Widgeon, Copain, Avalon and Flanders – and there it grew, waving in the wind on those open sweeping downs, a delight to the eye.

'Is there any disease in it?' I asked Mr Wookey.

He laughed and said, 'I don't know – and I'm not going to look. For if I found disease I couldn't do anything about it. For I am not *allowed* to spray!' He grinds most of this wheat in his own mill, and bakes much of the flour in his own bakery, and it is sold as organic produce.

Now I have on my table as I write this a recent copy of the *Irish Farmers' Journal*. This journal, which is like a thousand others throughout the world, gives farmers routine directions about how to farm their various crops for this month. Under the heading 'Winter Cereals' (meaning mostly wheat), this is part of what it says:

Routine disease spraying can begin with the coming week on advanced crops provided the weather is suitable. Rhyncho on new growth should also be treated. All second and subsequent crops should be sprayed for eyespot.

For broad spectrum control use Tilt, Radar, Sportak, Bayleton, Corbel, Mistral, Missile . . .

For eyespot and additional rhyncho control add MBC, e.g. Bavistin, Derosal, or use the proprietary mixtures Tilt MBC, Sportak Alpha, Impact MBC or Bayleton BM. In intensive cereal situations where MBC has failed to prevent lodging, some resistance may have developed. In such circumstances Sportak Alpha is preferred . . .

Early-sown wheat crops which have reached stem extension can get their first application of cycocel where the policy is to split . . .

Spray now for overwintered weeds . . . for chickweed and cleavers two pints an acre CMPP is cheap and effective. Britlox, Cleval, Starane 2, Ally, Springclene, Seloxone, Mylone and tank mixes of these with CMPP for broad spectrum control.

Arelon, Hythane, Dicurane (half-rate) and Twin Tak for meadow-grass, hoe grass; Avenge, Dicurane, Commando for wild oats.

One wonders if this is *the same world* as the farms of Mr and Mrs Wookey! There we stood looking at a great sweeping area of fine wheat, obviously disease-free, obviously growing very clean of weeds, and none of it had had one drop of any of those poisons to protect it and aid its growth. Not one drop and much of that land had not had one drop for a decade and more.

It is quite amazing that in practically every country I go to I find that orthodox, establishment scientists and advisers will do anything rather than admit that there could possibly be any virtue in non-chemical agriculture at all. Agricultural experts – ministers of state even – have walked over the Wookeys' farm, seen their wheat, seen their records and books, seen this example of a very successful and profitable enterprise that uses no chemicals, and they are like the country boy when he first saw a giraffe: they say 'I don't believe it!' Establishment agriculturalists will not believe their own eyes when it comes to looking at a good crop of wheat that has been grown with no chemicals.

Why are the Wookeys, and a growing number of farmers like them throughout the wheat-growing part of the world, able to grow perfectly good wheat without chemicals and yet the readers of the *Irish Farmers' Journal* and the readers of a thousand other journals like it apparently can't?

You will already have noticed that, under the Wookey rotation, the land has five years of ley (grass and clover) for only four years of corn growing (or white straw crop as the Americans would have it). Opponents of biological farming would say that this means that you only have a crop for four years out of nine and what is the good of that? For the orthodox cannot see that grass, too, is a crop. They have become so accustomed to monoculture that, if they classify a farm as a wheat farm, then they cannot accept the value of anything but wheat that is grown on it. Their philosophy has divorced the farm animals from the land. They cannot accept that it might be possible to bring them back again.

The Wookey lands stay in good heart because of all that grass and clover that is ploughed back into them – but also because of the fact that there are 560 head of beef store cattle on the farms; 1700 ewes and lambs; besides 40 horses.

Why horses? Because the Wookeys believe in mixed stocking, which is a good old farming concept. The animals of one species ingest the parasites voided by those of another and, in so doing, destroy them.

So, besides all that mass of clover and grass, all the dung goes into the Wookey land. Therefore the land grows fairly heavy crops of wheat without an ounce of artificial fertiliser. As for disease – crops grown on rich, biologically active land suffer very little from it.

Now, as we have heard, Liebig stated in the 1840s that you cannot continue to take necessary elements out of a piece of land without putting anything back. This is of course so. The flow of nutrients in orthodox agriculture may be described as linear. Nutrients come from the oil wells, and from potash and phosphate mines, are fed to crops, the crops (some going through animals first) are fed to humans, the resulting sewage (which contains virtually all of the nutrients) is dumped into the sea from which it is irretrievable. So the flow is in a line.

The flow of nutrients in biological agriculture however is cyclical. In theory, at least, the nutrients come from the soil, go through the plants, go, sometimes through animals, through humans, and then go back to the soil. In theory at least: all biological theorists postulate that one day human sewage must (after suitable composting) go back to the soil, but we all realise that there are severe customary and taboo barriers to this, apart from the fact that human sewage is at present mixed in with so many poisons from industry and inorganic agriculture that it is unsafe to put it on the land.

Inorganic agricultural theory is above all simplistic. It just cannot cope with the enormous complexity of interdependent living systems. It sounds so simple: analyse your soil for the three main plant needs – nitrogen, potassium and phosphate – add any deficiencies of these in a soluble form easily absorbable by the plant roots, and – hey presto – you have the perfect agriculture. We will all get rich.

But it is not as simple as that. In the first place, the organic content of the soil under such a regime diminishes because it is not renewed. This causes debility and disease among the crops. The simplistic answer to this is to hand the problem over to our brothers the chemists! This is done and the chemists produce poisons which kill the disease organisms and thus enable the crop plants to grow. This, of course, further destroys the living organisms of the soil and pushes the soil further on its way to becoming simply a sterile layer of powdered rock. Then erosion begins and the soil starts to go. This is only just beginning to happen on a serious scale in western Europe and no doubt we will see all kinds of fixes, in due course, to arrest it.

Then the disease organisms of crops and stock, able to hit much harder than previously because of the unnatural way these crops are grown, develop resistance to the poisons. So the chemists invent worse and stronger poisons. But no species of pathogen has yet been eliminated from this planet by poisons. These small pests and disease organisms seem to have an unlimited power of adapting.

We must realise that the world's stores of fossil fuel (for fixing atmospheric nitrogen), and potash and phosphate rock are not inexhaustible. The Arab oil-sheiks suddenly realise they are sitting on a fortune and the price of nitrogen fertiliser trebles in six months! Suddenly Mr Wookey finds his farming methods, which do not have need of nitrogenous fertiliser, to be considerably more competitive. Phosphates, particularly, are giving cause for concern to people who can see further than the day after tomorrow. Most of the world's available mineral phosphate has already been used up and dumped into the sea from where it is irrecoverable.

The average biological farmer would not be averse to seeing some of this mineral phosphate or potash being put on his fields, if he thought they needed it (as most of us will use lime if our soil is too acid), but he will not become dependent on artificially fixed nitrogen, and it is this latter which is really the crux of the whole business.

Inorganic farming is dependent, and becoming increasingly dependent, on ever more massive applications of nitrogen which has been 'fixed' from the free nitrogen in the air to turn it into a compound capable of being consumed by plants. Sir Kenneth Blaxter, Director of the Rowett Research Institute, Aberdeen told the Oxford Farming Conference recently that 'it does not seem to me a very good return on a sixteen-fold increase in draught power and a twenty-fold increase in nitrogen fertiliser input to obtain simply a doubling of production.' (The increase in draught power that he was referring to has come about because of the intractability of soils which are deficient in organic matter. They have to be literally torn apart by brute force.)

Nitrogen, in natural soils, is constantly 'fixed' from the air by bacteria. In any healthy organically-rich soil there are millions of nitrogen-fixing bacteria in every cubic inch and these creatures survive simply by their ability (shared by no higher forms of life except man in very recent years) to fix free atmospheric nitrogen into a nitrate compound which they – and in turn higher forms of life – can use. A very well-known and visible example of this nitrogen-fixing capacity of bacteria is in the nodule bacteria that affix themselves to the roots of members of the *leguminosae* family – the peas, bean and clover family. Anyone can see, with the naked eye, the small pimples or nodules on the roots of healthy plants of this family. These nodules are specially grown by the plants to house the bacteria that live in symbiosis with them: the plant providing shelter and other nutrients besides nitrogen – the bacteria supplying the plant with nitrogen.

If you dump large quantities of fixed soluble nitrogen on land you suppress the nitrogen-fixing bacteria. Any farmer knows, for example, that in a grass and clover mixture added nitrogen encourages the grass at the expense of the clover and the latter diminishes. If, added to this, you do not renew the organic content of the soil you take away an essential ingredient of the diet of nitrogen-fixing bacteria and this diminishes their numbers too.

So, fix nitrogen from the air with enormous consumption of fossil fuel power, dump it on the land, and you put the vast army of free and willing workers, that have always been used to supply the soil, and us, with fixed nitrogen, out of work. I cannot help seeing a close connection between this unemployed army of the soil with that other army of unemployed at the labour exchanges in Europe and throughout the world. The people who should be caring for the farm animals on the farms, carrying the straw to bed the animals, carrying and spreading the dung of the animals, composted with the straw, back to the land, and doing the other jobs that good husbandry requires, are all kicking their heels about in huge cities being supported in idleness (through no fault of their own) by their fellow men.

For biological farming, make no doubt about it, takes more labour than conventional farming. I remember the first farmer I worked for, Mr Catt, saying to me, 'Chemical farming is lazy man's farming!'

Also, in the short term at least, chemical farming can produce a little more of a given crop per acre than biological farming can. But this is only by massive chemical inputs, which can only get more expensive as the years go by. It is not a sustainable agriculture.

Biological agriculture does not mean going 'backwards' to a more primitive agriculture: not even to the extremely sound and sophisticated agriculture of people such as Mr Catt. Enormous strides have been made in this century in evolving new methods of biological agriculture. The research for this has had to be done by dedicated individuals and groups of individuals, until recently without any support whatever from government or official sources. Having been to the sort of agricultural college at which all our orthodox agricultural scientists and advisers were trained, I find it easy to understand the reductionist and simplistic attitude that most of them carry through life. Confined in their own very narrow discipline, they are unable to comprehend the vast complexity of the soil community which spreads right round our planet and comprises as its members every living thing. True, the establishment is beginning at least to notice the *existence* of the organic movement, and that is a change for the better. At the 1985 Oxford Farming Conference biological farming was discussed. It was, of course, damned out of hand. Mr Peter King, General Secretary of the Society of Chemical Industry said, 'Given the imperative need for low-cost production, organic farming will have only limited application. There will always be the sucker market in Hampstead but not a lot more.'

Professor Ronald Bell, Director-General of the British Government's Agricultural and Advisory Service, added, 'There were those whose instinct led them to suppose that "dog and stick" farming of yesteryear must be more "natural" and therefore better . . . however farmers would want to produce what the consumer preferred and was prepared to pay for.'

With such attitudes, it is not surprising that biological farmers have had absolutely no help whatever from establishment research institutions. And the new-wave biological agriculture desperately needs hard and serious scientific research. The problem with orthodox agricultural research, apart from the fact that it is largely financed by the chemical industry, is that it is reductionist. It splits every problem up into compartments and considers these in isolation. At my college the botany lecturer knew nothing about chemistry and the chemistry lecturer knew nothing about botany, and so on. All western science has gone through this reductionist phase but at last it shows signs of moving on. More and more scientists are discovering that a holistic view of the universe is the only one that really makes sense. There are signs that even agricultural scientists (in a few isolated cases so far – but they are increasing) are coming round to this position.

The biological thinker realises that many things must be taken into consideration in deciding any policy. An orthodox agricultural scientist in considering, say, what applications of nitrogen should be put on wheat, will conduct a few simple field-trials, find that the law of diminishing returns sets in after X number of units of nitrogen per acre, and therefore make X his recommendation. A biological scientist should bring in a thousand other considerations: the rate of unemployment and crime in the cities, the probable oil and natural gas reserves of this planet, the nutritional value of crops, the probable effects on soil degradation, erosion, disease, soil biology – you could fill a book with subject headings alone.

1968 and 1969 were unusually wet years in England and a great deal of damage was done to heavy soil all over the country by the action of weighty machinery on land which had been depleted of organic material. Even the farming press (usually on the side of the chemical and machinery industries) expressed concern. The Minister of Agriculture therefore instructed the Agricultural Advisory Council to conduct an inquiry into it. The branches of the National Agricultural Advisory Service all over the country contributed to this and a report was prepared which completely whitewashed existing agricultural practice. The very first words of the introduction of the report tell the whole story: 'Much is said and written nowadays in rather vague and sweeping terms about the way we farm our land.' In other words, the opposition was to be scornfully dismissed in the very first paragraph. To read the report – and there are over 100 pages of it – leaves one with the euphoric feeling that the whole problem was an illusion, the damage to the soil was only imaginary and all is well in the best of all possible worlds.

It is perhaps embarrassing to the establishment that, at the very time this report was being written, another establishment body, the Soil Survey of England and Wales, which comes under the Geological Survey Department, was carrying out investigations into soil erosion in England and Wales.

We visited the offices of the Soil Survey at Cambridge and we went to see the man chiefly concerned with erosion research, Mr R. Evans.

'I gather there is no soil erosion in England?' I said to him.

Mr Evans leapt from his chair. 'No soil erosion? No soil erosion? Here – look at this!'

His room was nearly filled with chests of shallow drawers. Out of these drawers he began to haul aerial photographs. They were all of the English countryside and they all showed gross soil erosion. Soon the floor was covered with them.

In East Anglia, Mr Evans told us, as much as eleven tons per acre (eighteen per hectare) of topsoil is being lost per year over large areas in Bedfordshire; on sandy-loams, annual losses have been found to be in the order of four to eighteen tons per acre (ten to forty-five per hectare); even on chalky soils, where it was thought erosion was negligible, it has been found that up to twenty-four tons per hectare is being lost. (For comparison it has been found that soil loss from woodland or heathland is about 0.04 tons per acre, or 0.1 per hectare. The greatest renewal capacity of soils has been found to be about one ton per hectare per annum.)

The *only* way to prevent erosion and soil degradation on arable land on this planet is to go forward (you can never go back) to ecologically sound agriculture – to biological farming, in fact. And all over the world, the small army of biological farmers and growers is growing. In spite of having no funding but their own hard-earned money for research, they are learning how to make a living, how to produce good and clean food for people to eat, free from poisons, low in nitrates and nitrites, and they are learning how to keep plants and animals in conditions not too far removed from the conditions for which they were evolved.

In the future we will have a varied and beautiful countryside again. Contented farm animals will be seen on every farm and the day of the swollen agribusinesses, with their vast featureless areas growing only one crop, laced with chemicals, will be over. We will see family farms again, and a human landscape, and trees and orchards and hedges. Wild birds and other creatures will come back and the rights of other creatures on this planet besides man will be recognised and understood. Villages will come alive again, and not just be dormitories for city commuters. They will be inhabited by that now nearly extinct breed – true countrymen and true countrywomen.

And the soil, that world-surrounding matrix that we all came out of and into which we shall return, will be treated with reverence and respect, and will recover the health and wholeness that God, and nature, intended it to have. For *all* life on this planet depends directly and entirely on the soil and to damage and destroy the soil is an affront to God.

The Amplification of Man
———— Herbert Girardet ————

In the plains of the lower Rhine at Hambach, near Aachen, Europe's latest opencast lignite mine began coal production in 1984. Here are deposits of the soft brown coal up to 100 yards thick. Twenty million years ago they were lush tropical forests. These were quite literally swallowed down as the land subsided over millennia; more forests grew and they, too, were sucked underground. Finally, the thick layer of decayed trees and vegetation was covered over by silt and gravel brought down by the Rhine from the glaciers of the Alps, 400 miles to the south.

At Hambach, we saw five giant bucket-wheel excavators at work. In order to get at the coal, a layer of overburden up to 300 yards thick has first to be removed. The huge hole in the ground – it will soon extend to forty-two square miles (110 square kilometres) – will yield 2500 million tons of lignite over the next fifty years. Next to the hole, a hill is being built which will soon be 150 yards high. This is being planted with bushes and trees.

From the hill one can see the sheer scale of the operation. The 'hole of the century' is already 200 yards deep and will eventually reach over 500 yards down. It hums with the vibrations of electric motors, driving excavators, conveyor belts and powerful electric pumps. Conveyors stretching for many miles transport the overburden from the hole to the side of the new hill. Down below, the coal is being dug away. It is transported to nearby power-stations which produce nearly a quarter of West Germany's electricity.

The village of Niederzier on the edge of the mine is about to be 'transferred' to another location. Farms are being re-established on land reclaimed from former opencast mines. Forests, too, are given new locations on the hills that are now growing out of the previously flat landscape.

In half a century the lignite deposits at Hambach will have gone up in smoke. There are more deposits nearby, but they are even deeper, even more difficult to mine.

Coal mining in the Rhineland first started 200 years ago. The lignite was just below the surface and could be dug up by men using picks and shovels. After 1945, the first deep opencast mines were developed. Bucket-wheel excavators came into use, replacing small-scale mechanised equipment used by the miners. Today, at Hambach, five men operate one excavator that can dig up 240,000 cubic yards of overburden or coal a day. The machines are 105 yards high and 250 yards long; weighing 13,000 tons each they are the heaviest machines on the surface of the Earth.

In the mine the pumps have to be kept running day and night to remove the groundwater. A total of 7040 gallons, or 32,000 litres, are pumped out every minute, 300 million cubic yards per year, from 850 wells. Eventually the water-table will have to be lowered by over 500 yards to keep the mine dry.

These figures illustrate the scale of contemporary mining operations, an activity which has become an essential part of human identity. Without fossil fuels we would be different creatures. The harnessing of fire in the steam engine, the steam turbine and the internal combustion engine put at our disposal a motive force which has transformed human impact on this planet. A man with a pick and shovel is a different creature from a man who operates a bucket-wheel excavator. Standing next to each other without their equipment they may look the same but in ecological terms they are as different as a mouse and a dinosaur.

Hunter-gatherers or even self-sufficient peasant farmers are creatures whose impact on the planet is purely biological; they consume only plant material and meat whose chemical constituents are returned to the soil. Their metabolism is little different from their animal cousins.

Even medieval knights on horseback were largely biological creatures, though their ecological impact was greatly enhanced by their animals' need for grazing land.

Technological man, on the other hand, cannot be defined in purely biological terms. We consume coal, oil, uranium, iron and aluminium as part of our metabolism. We breathe out sulphur dioxide and nitrogen oxides from the chimneys of our power-stations. The heavy metals deposited in the soil on the edge of our motorways or in the mud at the bottom of our rivers are part of man's secretions.

The bucket-wheel excavator that digs up coal in the Rhineland is an extension of our amplified being. The crop-spraying aeroplane that sweeps across a wheatfield in East Anglia is part of our ecological impact. The drilling-rig in the North Sea or the supertanker on its way from Karg Island to Rotterdam are both aspects of the ecological identity of the 'amplified man'.

Man does not only eat food today, we also eat land: every ounce of soil that is washed down from a denuded hillside is part of our consumption pattern. Every area of forest or grazing land that is converted to desert is an integral part of our metabolism.

In 1980, every American consumed nearly 26,460 lbs, or 12,000 kilograms, of 'coal equivalent'. West Germans about half that amount, Kenyans used 440 lbs and Ethiopians consumed 55 lbs per capita. An

American uses a thousand times more oil than a citizen of Ruanda. In 1975, each American used metals valued at 200 dollars, a western European 120 dollars, an African 4 dollars. These figures do not only reflect different living standards but also different scales of ecological impact.

In 1982, there were 385 cars for every 1000 West Germans, fifteen per 1000 Turks. If India wanted the same number of cars per inhabitant as the USA it would have 128 million, 237 times more than today. If all countries in the world were to reach the production levels of the USA or West Germany, the global ecological consequences would be ruinous.

The overall demands of an American throughout his lifetime are sixty times greater than those of an Indian. An Indian farmer produces sixteen calories of food energy for every calorie of technical energy used in the farming system. In the USA for every calorie of energy used in the farming system only one-half of a calorie of food energy is produced.

Americans consume nearly a ton of grain per capita per year, Africans eat one eighth that amount. Every German has as much land at his disposal in another country to yield food for him as he has in his own country. Cassava is imported from Thailand, maize from the USA, groundnuts from Niger where hunger and starvation are rife – most of this is used for feed in pig and chicken units. In Africa maize is a staple crop for humans, in Europe and the USA it is almost exclusively used as animal feed. Ten times as much land is required to produce one calorie of food energy contained in beef as that contained in bread.

In the USA the average citizen now consumes 240 lbs, or 110 kilograms, of meat per year. A UK citizen eats 165 lbs, a Russian 112 lbs, a Brazilian 70, a Chinese 46, a Nigerian 13, an Indian 2.4. Around 40 per cent of world cereal production is used to feed livestock and in the richer countries this figure can be as high as 75 per cent. In fact, in the USA 90 per cent of cereals used in the home market go into animal feed.

It is estimated that 75 billion tons of topsoil are at present lost annually throughout the world through wind and water erosion, fifteen tons for every human inhabitant of this planet. The mountain slopes of Nepal alone are losing a quarter of a million tons of soil a year; downriver in the Bay of Bengal, soil is accumulating which, when it breaks the surface of the water, will create an island of around 12.3 million acres (5 million hectares).

In the USA during the dustbowl years in the 1930s, 98 million acres (40 million hectares) of land were severely damaged. Today, under the pressure of technological farming systems, one-third of US croplands – 123 million acres – is undergoing a marked decline in long-term productivity owing to soil erosion. Ethiopia which is only one-sixth the size of the USA – is thought to be losing as much topsoil as the whole of the USA. Globally around twenty-seven million acres, or eleven million hectares a year are lost through erosion, desert encroachment, toxification and cropland conversion to non-agricultural uses. By the year 2000, 18 per cent of the world's arable land could be lost to farming if present trends continue.

All these figures give some impression of the immense scale of human impact on our planet today. The ecological effects of poverty and of affluence are closely intertwined. Rich and poor countries are doing severe damage to forests and to farmland. Everywhere in the Third World cities are growing as populations expand and rural life loses its viability. By the turn of the century half the population of the Third World is expected to be living in cities. They will be an enormous drain on the fertility of the farmland and pastureland which will have to feed them. In addition, the export of cash crops is a further major loss of plant nutrients from the soil. And all too often the poor countries will not be able to afford to buy the fertilisers needed to replace that fertility. Very few countries are likely to follow the Chinese example of waste-recycling in which urban waste and sewage is returned to farming areas.

About half the world's population has firewood and animal dung as its only source of fuel. The situation was made worse by the rapid increases in oil prices in the 1970s, when many millions of people in poor countries suddenly could no longer afford to buy kerosene or fuel oil. Today, deforestation is perhaps the most worrying aspect of our onslaught on the planet. Every month, an area of forest the size of Wales is felled. Where they are not under attack from the ravages of air pollution and acid rain, forests are pushed back by axes, chain-saws, bulldozers or fire. In the Third World, traditional forest farmers, the so-called slash-and-burn agriculturalists, are forced to intensify their land use owing to lack of adequate farmland. Throughout the tropics, forests are being transformed by logging, farming, ranching and mining. There will be very little tropical forest left in West Africa, Middle America and tropical Asia by the turn of the century. The great rain forests of the Congo Basin in Zaïre and of the Amazon Basin in Brazil will be much reduced in size.

The amplification of man today is taking place both through the increased use of technology and through population growth. There are 5 billion of us now and there will be 6 billion by the end of the century, the end of the millennium. Sheer pressure of numbers is part of the problem of Third World poverty. Ecological breakdown and destruction of soil are serious problems in many countries, but so often they result from land shortages due to lack of land reform. Throughout Latin America, for instance, poverty has resulted from the concentration of huge areas of farmland in the hands of a tiny minority.

In Brazil, the construction of the Transamazonian Highway was undertaken by the government of President Medici in 1971 in order to give landless people in the poor north-east of the country access to farmland. Medici simply followed the pattern of development set by his European forefathers: the removal of forests is a precondition for the creation of farmland. On both sides of the new road farmers were settled. They burned the forests and planted crops in the ashes. For one or two years the farms flourished, but on most settlements the land would not yield a third harvest.

The authorities had ignored one vital fact: nine-tenths of the nutrients in the tropical rain forests are contained in the trees themselves. Only one-tenth is held in the soil. Whenever a leaf falls off a tree it is almost immediately decomposed. A deep leaf litter does not build up as in a temperate forest. A humus-rich layer of soil does not form.

Settlement of farmers along the Transamazonian Highway proved a failure. But the road was cut deeper and further into the forest – the greatest forest on earth, which extended to 690 million acres (280 million hectares) a few decades ago. The cutting edge of the invading civilisation is men with chain-saws and bulldozers, felling trees, pushing the stumps aside, levelling the ground. The roads are cut in endless straight lines. In the late 1970s they became the access routes for the 'new farmers', multi-national companies who were into beef ranching. They were given concessions of forest land which could be the size of a country like Luxemburg. Half of these were to be kept as forest under government rules, the other half could be turned into ranchland for beef cattle.

The removal of the forest was accomplished with military efficiency. Defoliant sprays, first tried out in the Vietnam war, were sprayed on an area of forest, maybe 250,000 acres, about 100,000 hectares at a time. When the trees had dried up sufficiently to be inflammable – some were cut down to start the bonfire – a section of forest was set alight with old lorry tyres and oil. In some instances napalm bombs were used. Some of the biggest fires ever seen on Earth were started in this way. They were clearly visible on satellite photographs. Aeroplanes were forced to fly around and above the huge plumes of smoke which could reach up thousands of yards.

The luxuriant forest, harbouring innumerable species of plants and animals, was thus reduced to a layer of smouldering ash, with scarred tree-stumps sticking out of the ground. Aeroplanes scattered seeds of African savannah grass on to the naked soil. In forty-five days they could grow waist-high. If the forest tried to regrow from seeds in the ground the grass was burned off. Then Zebu cattle were introduced to feed on the newly established grass.

There is little employment on the ranches themselves, one man is expected to look after 1000 cattle. There are more jobs in the slaughterhouses. Cattle are usually killed when they are four years old. Their carcasses are loaded on to aircraft and they are deep-frozen high up in the atmosphere on their way to a freezer warehouse in Florida or Frankfurt. Production costs are lower on Amazonian ranches than in Europe or North America; huge profits can be made even if the soil produces grass – and beef cattle – for only a few years.

Now the Brazilian government no longer encourages new ranching enterprises in Amazonia. Burning down large areas of forest is perhaps not the best way to use valuable hardwood resources. Enough roads have been cut into the forest to make logging a viable operation. And 'oil' has now been discovered in Amazonia. Brazil does not have oil of its own and it cannot afford to import much. An increasing number of cars are powered by *alcohol*. In 1981, 250,000 cars were produced in Brazil which can run exclusively on alcohol. Several million acres of fertile land in Brazil are under sugar-beet plantations from which the alcohol is distilled. (This land is, of course, not available to landless peasants.) Now logging operations in Amazonia can be developed simultaneously with production facilities for fuel alcohol. The valuable hardwood trees are used for export, softwoods can be distilled into wood alcohol.

Meanwhile, mining operations and associated hydro-electric schemes are coming on-stream. In north-eastern Brazil, the Tucurui Dam, one of the world's largest reservoirs, is now going into operation. It will power the biggest open-cast mining operation on the planet. In Guyana, the Upper Mazaruni Dam is providing electricity for bauxite mining and processing. Vast areas of rain forest have been flooded with trees left to rot in the rising water.

The forest people are left stunned by the onslaught. They say that the forest can provide everything a man needs. The settlers on the other hand – be they ranchers, road builders or mining engineers – need to *remove* the forest to make their living. The Amerindians have lived in the Amazon forest for tens of thousands of years and their survival proves its viability for human habitation. The Amerindians are deeply offended by the white man's attitude to the forest as a 'green hell'. Their intimate familiarity with its immense variety of plants and animals makes them the true experts of the Amazon forest. They, too, are farmers but they cut small clearings into the jungle. They plant their crops in the jungle gardens, a great variety of maize, beans, cassava, taro, tomatoes, and vegetable plants virtually unknown to outsiders. They abandon the gardens after two or three years when the soil fertility is exhausted, but they leave banana, avocado and mango trees behind, to which they can return to harvest the fruit until the forest has reclaimed the land with larger trees.

Forest people, as we saw at the beginning of this book, are immensely knowledgeable about their habitat. The Amazon forest is among the most diverse environments anywhere on this planet. On one hectare of rain forest in Brazil as many as 400 different kinds of trees have been found. The Amerindians are the guardians of the greatest botanical treasure-house on Earth, containing many plant species not yet recorded or named by science – that is, our science.

Most settlers of Amazonia regard that which is above ground as a nuisance to be cleared away. The treasures they are after are hidden in the ground: gold, diamonds and all kinds of precious stones. Many thousands of prospectors are spurred on by the dream of that nugget of gold the size of a fist or the diamond the size of a golf ball.

Amazonia is the last great frontier to be crossed in man's great quest to conquer all of the Earth. The magnificent wilderness must be *tamed*. Tens of thousands of years ago, man came naked, equipped only with stone axes, bows and arrows and blow-pipes. Today, we come kitted out with bulldozers, aeroplanes and floating pulp-mills, built in Japanese shipyards and towed across the ocean to be anchored on the banks of the Amazon. How many trees must be cut down to make such an operation worth-while?

At Manaus, the jungle town where Europeans built their great opera house at the time of the rubber boom, scientists in INPA, the Institute for Amazonian Research, are frantically classifying plant and animal species previously unknown to science. Many are passionately committed to preventing the final holocaust of the forest and its people. They know that they have only a very little time to prove that it is *worth* preserving the forest, or most of what is left of it.

INPA scientists acknowledge the essential contribution made by Amerindians to an understanding of the ecological complexities of the forest environment. The knowledge of the forest people is not only confined to edible plants that grow in Amazonia. They have developed a great variety of plant-based remedies against infections, parasites, internal diseases, insect bites, cuts and bruises. Unfortunately they did not have any medicines for the treatment of unfamiliar viral diseases such as measles and influenza which were introduced by the white man and which proved deadly in many instances.

Amerindians also 'designed' composite arrow poisons such as curare and its many derivatives, which are used in hunting: the poison on the arrow tip kills animals but does not affect the people who cook the meat and eat it. In contemporary medicine, an alkaloid derived from curare has become an important 'chemical tool'. It is widely used in open heart surgery.

The Amerindian tribes also developed plant-based contraceptives. Some of these are being analysed by biochemists in Brazil, Europe and North America. The London-based ethnobotanist and chemist, Dr Conrad Gorinski, has evidence that herbal contraceptives extracted from forest plants contributed to the traditional population stability of the Amerindian tribes.

Apart from a small number of scientists – botanists, chemists, zoologists and anthropologists – the world has no time for 'naked savages'. They are in the way of progress: of roads, ranches, plantations, reservoirs and mines. As the forest is opened up by road construction and the removal of trees, contact with the forest dwellers is inevitable. All too often this has resulted in the destruction or drastic reduction of tribes. From 1900 to 1980 the Amerindian population in Brazil had gone down from one million to around 200,000. Some tribes have been airlifted out of 'their part of the forest' into remoter parts of Amazonia. But they find it difficult to adjust to unfamiliar trees and forest plants in a strange new environment.

The tropical moist forests of South America, Africa, Asia and Australia extend to about 2225 million acres (900 million hectares); which is 6 per cent of the world's land surface. They are thought to harbour about three-quarters of the estimated 10 million plant and animal species that exist worldwide. The destruction of tropical moist forests is proceeding at an unprecedented rate: by the mid-1970s, 40 per cent had been lost according to a UN Food and Agriculture Organisation report. The US Academy of Sciences estimated in 1980 that up to 50 million acres of tropical moist forest were being destroyed or degraded annually. Large-scale deforestation usually results in permanent wasteland, since trees cannot re-establish themselves on the parched and nutrient-poor soil.

The multilayered canopy of an undisturbed rain forest has been called a 'pasture on stilts' as it provides sustenance for a large range of monkeys, birds and, of course, insects. The shade, and the moisture which circulates in the forest, keeps the temperature on the forest floor down to about 27 degrees Celsius. Once the trees are removed, the soil temperature rises to 45 Celsius or more, and the bare ground reflects the sunlight back into the atmosphere. Large-scale deforestation causes a so-called albedo effect: the incoming solar energy is not utilised by plant growth but contributes to a rise in air temperature instead. Forest destruction induces air turbulence and leads to erratic weather patterns. Torrential rain can no longer be absorbed by the 'living sponge' of forest vegetation. The water runs off the bare soil and swells streams and rivers. In Manaus, flooding has become a great problem in recent years. This is thought to be caused by loss of tree cover in nearby forest areas.

The impact of forest removal in the tropics on the climate is a growing cause for concern. A large-scale loss of moisture from the tropical belt due to deforestation, and a rise in surface and air temperature as a result of loss of vegetation cover, can lead to self-accelerating desertification. Worsening conditions for agriculture on the southern edge of the Sahara have been associated with deforestation in west Africa. It has been suggested that forest destruction in Middle and South America may contribute to a reduction in rainfall as far north as the southern states of the USA.

The forests of the world no longer absorb carbon dioxide, they actually release it. Two-thirds of the wood consumed every year is burned as firewood. This, together with forest fires such as those still flaring up in Amazonia, is thought to release as much carbon dioxide into the atmosphere as all the fossil fuels we burn. A steady increase in carbon dioxide concentrations in the atmosphere has been observed at the Mauna Loa research station in Hawaii, where measurements have been taken since 1957. In the last thirty years the carbon dioxide content of the atmosphere has increased by about 10 per cent. Carbon dioxide causes solar radiation to be retained within the atmosphere; a steady increase in carbon dioxide concentrations as a result of wood and fossil fuel combustion is predicted to lead to an increase of air temperatures of several degrees Celsius some time in the next century. The consequences of this 'greenhouse effect' for the global climate could be staggering. Predictions range from drought conditions in the wheat belt of the USA and the Soviet Union to a major reduction in ice cover at the North and South Poles with a resulting rise in sea levels.

Today, the consequences of the power of man to affect the biosphere are there for all to see. We may prefer not to spend our holidays travelling around in eroded and depleted landscapes, though those of us who have been to countries bordering the Mediterranean have had first-hand experience of the impact of earlier civilisations on their 'host landscape'. Television cameras bring evidence to our living-rooms from further afield. It is clear that we shall continue to run down the living resources of this planet at our peril.

The *remains* of tropical forests stored away under the Earth's crust in the form of lignite, coal and oil deposits are being used up at an exorbitant rate. *Living* tropical forests will have all but disappeared in fifty years if the present rate of exploitation continues unchanged. We are wasting soil at a rate vastly exceeding its capacity to renew itself. Air pollution originating from our chimneys and exhaust-pipes is causing a dramatic deterioration in the condition of forests in the temperate zones. Precious plant nutrients are not being recycled but end up as pollutants in watercourses.

The amplification of man has been achieved at a tremendous cost to the global environment. And yet, hunger and misery are affecting more people than ever before. *We will improve things only if we are prepared to share out the resources of our planet among ourselves, between ourselves and future generations and between mankind and the living world.* The rich countries can no longer afford to behave as if this were a clearance sale of global resources.

To live on this planet we have to understand what we need to do to keep it habitable. Such an understanding is gradually emerging in the minds of many people who have become tired of the hectic quest for material progress which has determined our actions in the last few decades.

Today, we can assess the sprawl of deserts from cameras mounted on satellites. We can study the depletion of soil life through the eye-pieces of microscopes. We can verify the depletion of fish stocks in the oceans with the help of computerised scanners. Are we going to act on the available evidence?

Ecology is the science of planetary housekeeping. We cannot afford to ignore its finding. To understand the ecological impact of modern, 'progressive' societies, we need to take account of their total metabolism, which includes the fuels we burn, the raw materials we put through our machines, the forests we consume and the waste we dump. Ecology is *long-term* economy. Running down the resources of the planet is bad economics. Only a culture that *enhances* Nature in its many varied forms can expect to have a future.

The amplified man, now equipped with a huge array of machinery, has the power to change the face of the Earth. So far we have found it easier to turn land covered with lush vegetation into wasteland than to do the opposite. Who was it who said that man's footsteps on the Earth are deserts?

First, the amplified man came on horseback, with iron swords and suits of armour. Then he conquered far-away places with sailing ships and cannons. Today, he has tanks and bulldozers, nuclear submarines and drilling-rigs at his disposal to assert his will.

Where technology rules supreme biology tends to fall by the wayside. For a while we can ignore the environmental costs of our industrial production systems. But, as we have seen, we do not seem to be getting away with it.

There is a fundamental error in our present assumption that we can be masters of nature. It will never allow this. Long before we have completed the task of subduing nature we shall have removed the basis for our continued future existence.

Bibliography

Abel, Wilhelm, *Agricultural Fluctuations in Europe from the 13th to the 20th Century*, Methuen, London, 1980

Barney, Gerald O., *Global 2000*, Penguin, London, 1982

Berry, Wendell, *The Unsettling of America*, Avon, 1977

Bishop, Morris, *The Pelican Book of the Middle Ages*, Penguin, London, 1968

Bolsche, Jochen (ed.), *Was die Erde befällt*, Rowolt Verlag, Hamburg, 1984

Breuer, Georg, *Air in Danger*, Cambridge University Press, London, 1980

Bull, David, *A Growing Problem*, Oxfam Public Affairs Unit, Oxford, 1982

Burgwyn, Diana, *Salzburg, a Portrait*, Alfred Winter Edition, Salzburg, 1982

Camp, Wesley D., *Roots of Western Civilisation from Ancient Times to 1715*, John Wiley and Sons, New York, 1983

Carter, Vernon Gill and Dale, Tom, *Topsoil and Civilisation*, University of Oklahoma Press, Norman, 1974

Caufield, Catherine, *In the Rain Forest*, Heinemann, London, 1985

Caufield, Catherine, *Tropical Forests*, Earthscan Publications, London, 1980

Cipolla, Carlo M., *Before the Industrial Revolution: European Society and Economy 1000–1700*, Methuen, London, 1981

Darwin, Charles, *The Formation of Vegetable Mould through the Action of Earthworms, with Observations on their Habits*, London, 1881

Davidson, Basil, *Africa, History of a Continent*, Weidenfeld and Nicolson, London, 1966

DeBach, Paul, *Biological Control by Natural Enemies*, Cambridge University Press, 1974

Diercks, Rolf, *Alternativen im Landbau*, Ulmer Verlag, Stuttgart, 1983

Dudley, Nigel, Barret, Mark and Baldoack, David, *The Acid Rain Controversy*, Earth Resources Research, London, 1985

Duncan-Jones, Richard, *The Economy of the Roman Empire*, Cambridge University Press, London, 1978

Elsworth, Steve, *Acid Rain*, Pluto Press, London, 1984

Fernandes, Erick, Oktingati, A. and Maghembe, J., *The Chagga Home Gardens: a Multi-storeyed Agro-forestry Cropping System on Mount*

Kilimanjaro, Institute for Agro-Forestry, ICRAF, Nairobi, 1984

Der Fischer Weltalmanach, Fischer Verlag, Frankfurt, 1984

Focus on Rheinbraun, Rheinische Braunkohlenwerke, Koîn, 1982

Francé, Raoul, *Das Leben im Boden, Das Edaphon*, Edition Siebeneicher, Munich, 1975

Frayn, Joan M., *Subsistence Farming in Roman Italy*, Centaur Press, London, 1979

Galeano, *Die offenen Adern Lateinamerikas*, Peter Hammer Verlag, Wuppertal, 1974

George, Susan, *How the Other Half Dies: the Real Reason for World Hunger*, Penguin, London, 1976

Gimpel, Joan, *The Medieval Machine: the Industrial Revolution of the Middle Ages*, Victor Gollancz, London, 1977

Goldsmith, Edward and Hildyard, Nicholas, *The Social and Environmental Effects of Large Dams*, Vol.1, Wadebridge Ecological Centre, Cornwall

Gorinski, Dr Conrad, 'In Defence of Forest People', *Undercurrents*, no. 47, London, 1981

Goudie, Andrew, *The Human Impact*, Basil Blackwell, Oxford, 1981

Grant, Michael, *History of Rome*, Faber and Faber, London, 1979

Hancock, Graham, *The Challenge of Hunger*, Victor Gollancz, London, 1985

Herodotus, *The Histories*, Penguin, London, 1954

Hesiod, *Works and Days*, trs. Dorothea Wender, Penguin, London, 1973

Hopkins, Keith, *Conquerors and Slaves: Sociological Studies in Roman History*, Cambridge University Press, London, 1978

Hyams, Edward, *Soil and Civilisation*, 1952

Jackson, Wes, *New Roots for Agriculture*, Friends of the Earth

Koch, Egmont and Vahrenholt, Fritz, *Die Lage der Nation*, Geo Bücher, Hamburg, 1983

Kramer, Samuel Noah, *The Sumerians, their History, Culture and Character*, University of Chicago Press, Chicago and London, 1963

Lévi-Strauss, Claude, *The Savage Mind*, Weidenfeld and Nicolson, London, 1966

Liebig, Justus von, *Die Chemie in ihrer Anwendung auf Agricultur und Physiologie*, Vieweg, Braunschweig, 1865

MacMullen, Ramsay, *Roman Social Relations 50 BC to AD 284*, Yale University Press, New Haven and London, 1974

Mader, Hans-Joachim, quoted in *Natur und Landschaft*, no. 12, 1984

Mansholt, Sicco, *The Common Agriculture Policy, Some New Thinking*, Soil Association, Stowmarket, 1979

Masao, Fidelis T., *The Irrigation System in Uchagga: an Ethno-Historical*

215

Approach, Tanzania Books and Records, no. 75, 1974

Meadows, Dennis *et al.*, *Limits to Growth*, Angus and Robertson, London, 1972

Myers, Norman (ed.), *The Primary Source: Tropical Forests and our Future*, W. W. Norton, New York and London, 1984

Passmore, John, *Man's Responsibility for Nature*, Duckworth, London, 1974

Plato, *Critias*, trs by R. G. Bury, Heinemann, London, 1929

Preuschen, Prof. Gerhardt, *Die Wiederbelebung zerstörter Oekosysteme*, 1984

Quarto, Alfred, article in *Resurgence*, no. 107, Hartland, 1984

Raven, Susan, *Rome in Africa*, Longman, London, 1984

Sahlins, Marshall, *Stone Age Economics*, Tavistock Publications, London, 1974

Schütt, Prof. Peter (ed.), *Der Wald stirbt an* Streb, C. Bertelsmann Verlag, Munich, 1984

Snider, Gary, 'Wild, Sacred, Good Land', *Resurgence*, no. 98, Hartland, 1983

Stewart, Dr V. 'Soil Structure and Soil Testing', paper given at Aberystwyth University

Strahm, Rudolph, *Überentwicklung-Unterentwicklung, Stichwörter zur Entwicklungspolitik*, Burckhardthaus-Laetare Verlag, Gelnhausen, 1980

Todd, Michael, *The Walls of Rome*, Paul Elek, London, 1978

Turnbull, Colin, *The Forest People*, Jonathan Cape, London, 1974

Virgil, *The Aeneid*, trs. W. J. Jackson Knight, Penguin, London, 1956

Walters, A. Harry, *Ecology, Food and Civilisation: an Ecological History of Human Society*, Charles Knight, London, 1973

Washington, George, *The Letters of George Washington*, Doubleday, New York, 1966

White, Lynn, *Medieval Technology and Social Change*, Oxford University Press, London, 1962

Whitehouse, Ruth, *The First Cities*, Phaidon, Oxford, 1977

Woolley, Sir Leonard, *Ur of the Chaldees*, The Herbert Press, London, 1982